THEORIES

OF DEVELOPMENT

WILLIAM C. CRAIN

The City College of the
City University of New York

THEORIES
OF DEVELOPMENT
Concepts
and Applications

Prentice-Hall, Inc.
Englewood Cliffs, New Jersey 07632

Library of Congress Cataloging in Publication Data

Crain, William C (date)
    Theories of development.

    Bibliography: p. 267
    Includes index.
    1. Development psychology,   I. Title.
BF713.C72   155   79-16521
ISBN   0-13-913566-9

Editorial/production supervision and interior
    design by Penny Linskey and Linda Schuman
Jacket design by RL Communications
Manufacturing buyer: Edmund W. Leone

Printed in the United States of America

10 9 8 7 6 5 4 3 2 1

PRENTICE-HALL INTERNATIONAL, INC., *London*
PRENTICE-HALL OF AUSTRALIA PTY. LIMITED, *Sydney*
PRENTICE-HALL OF CANADA, LTD., *Toronto*
PRENTICE-HALL OF INDIA PRIVATE LIMITED, *New Delhi*
PRENTICE-HALL OF JAPAN, INC., *Tokyo*
PRENTICE-HALL OF SOUTHEAST ASIA PTE. LTD., *Singapore*
WHITEHALL BOOKS LIMITED, *Wellington, New Zealand*

**To Ellen
and to Adam and Tom**

Contents

Acknowledgments

This book grew out of discussions with my undergraduates at the City College of New York. My first thanks, then, are to them—for their interest and enthusiasm. I also owe a great deal to my own teachers, including Robert White, Erik Erikson, George W. Goethals, Bernice Neugarten, Wilbur Hass, Daniel G. Freedman, and Lawrence Kohlberg. My debt to Kohlberg in this work is particularly great; I have frequently followed his way of conceptualizing theoretical issues.

I would also like to express special thanks to my wife Ellen. Although very busy at home and at work as a pediatrician, she nevertheless helped enormously with every phase of this project. I also am grateful to my brother Stephen, who helped write the chapter on Chomsky, and to my parents, for their early intellectual stimulation. And I never would have written this book had it not been for the inspiration of my children, Adam and Tom. It was by watching them that I really began to appreciate the remarkable nature of the growth process.

The manuscript was substantially improved by Robert Liebert, Richard Rosinsky, Margaret Schadler, Douglas Kimmel, and William King, who read all or parts of it and made many useful suggestions.

I am grateful, finally, to those who gave permission to quote from various sources: W. W. Norton & Co., Inc. and The Hogarth Press Ltd. granted permission to quote from Erik H. Erikson's *Childhood and Society,* 2nd ed., 1963; Macmillan Publishing Co., Inc. and Bruno Bettelheim granted permission to quote from his book, *The Empty Fortress,* copyright © 1967 by Bruno Bettelheim; Little, Brown & Co. granted permission to quote the first stanza of "Growth of Man like growth of Nature" from *Poems by Emily Dickinson,* edited by Martha Dickinson Bianchi and Alfred Leete Hampson, copyright 1929 by Martha Dickinson Bianchi, © 1957 by Mary L. Hampson; and *Family Circle Magazine* granted permission to quote from Louise B. Ames's "Don't push your preschooler," in the December, 1971 issue, © 1971, The Family Circle, Inc., all rights reserved. Credit for the use of illustrations and other material is given within the text.

THEORIES

OF DEVELOPMENT

Growth of Man like growth
Of Nature
Gravitates within,
Atmosphere and sun confirm it
But it stirs alone.

Emily Dickinson

Introduction

We all have assumptions about the nature of development. We commonly assume, for example, that children's development is in our hands—that children become what we make them. We think it is our job to teach them, to correct their mistakes, to provide good models, and to motivate them to learn.

Such a view is reasonable enough, and it is shared by many psychologists—by those called learning theorists and by many others as well. Psychologists use more scientific language, but they too assume that parents, teachers, and others structure the child's thought and behavior. When they see a child engaging in a new bit of behavior, their first guess is that it has been taught. If, for example, a two-year-old girl shows an intense interest in putting objects into place, they assume that someone taught her to do this. For she is a product of her social environment.

There is, however, another tradition in psychology—a line of theorists dating from Rousseau—which looks at development quite differently. These writers, the developmentalists, are less impressed by our efforts to teach or otherwise influence children. Instead, they are more interested in how children grow and learn on their own. The developmentalists would wonder if this two-year-old's interest in ordering objects might not be a spontaneous one—something she has begun entirely by herself. Her concern for order might even be greater than that of those around her. For just as children, at a certain stage, develop an inner urge to stand and walk, they may also develop a spontaneous need to find order in their environment.

If we follow a child around, taking the time to observe the child's natural tendencies, we find that the child has many spontaneous interests. A one-year-old may become fascinated by a ball, a puddle of water, or a mound of sand—things which can be touched, felt, and acted upon. The child may examine and play with such objects for long periods of time. Such interests may be so intense and so different from our own that it is unlikely that they are the product of adult teachings. Rather, the developmentalists think, children may have an inner need to seek out certain kinds of experiences and activities at certain times in life.

The developmentalists—theorists such as Rousseau, Montessori, Gesell, and Piaget—do not agree on every point, and they have studied different aspects of development. Nevertheless, they share a fundamental orientation, which includes this interest in inner growth and spontaneous learning.

The developmentalists' concerns have been practical as well as theoretical. Montessori, for example, became dissatisfied with customary educational methods, in which teachers try to direct children's learning by rewarding their correct answers and by criticizing their mistakes. This practice, she thought, undermines children's independence, for children soon turn to the teacher, an external authority, to see if they are right. Instead, she tried to show that if we observe children's spontaneous interests, we can help provide tasks on which they will work independently and with the greatest concentration, without external direction or motivation. For there is an inner force which prompts children to perfect their capacities at each developmental stage.

In many other ways, the developmentalists have contributed to a new understanding of childhood and later development as well. Unfortunately, however, their writings have not received the full consideration they deserve. It seems that their emphasis on inner development has often struck psychologists as too romantic or too radical. Piaget, to be sure, has found a wide audience, but even he was ignored for decades.

There is one place where the developmentalists' concerns are seriously addressed. This is in modern humanistic psychology. Humanists such as Maslow have drawn heavily upon developmental ideas. However, the humanists have done this in a very implicit way, without recognizing how much they owe to earlier developmental contributions.

This book, then, is devoted to an appreciation of some of the outstanding developmental theorists. We will discuss some of the theorists who have followed closely in the footsteps of Rousseau, along with other theorists, including ethologists and psychoanalysts, who share a developmental outlook. We will discuss their concepts and some of the practical implications of their work. We also will review the first orientation we mentioned—that of the learning theorists, who help us understand behavior from a more environmental perspective. We will not cover learning theory in the depth it deserves, for this short book is primarily concerned with the developmental tradition. But we will try to get a flavor of their ideas. Finally, in the Conclusion, we will discuss the ways in which both the developmentalists and the learning theorists have been working in the humanistic tradition for some time.

Early Theories:
Preformationism, Locke,
and Rousseau

The two great pioneers in child psychology were Locke and Rousseau. Locke was the father of environmentalism and learning theory; his heirs are scientists such as Pavlov and B. F. Skinner. Rousseau began the developmental tradition in psychology; his followers include Gesell, Montessori, and Piaget. Both Locke and Rousseau made radical departures from an earlier outlook called preformationism.

PREFORMATIONISM

For centuries, people seem to have believed that children come into the world as ready-made miniature adults. As Ariès (1960) has shown, this view was predominant during the Middle Ages. Until the fourteenth century, paintings invariably portrayed children with adult body proportions and facial characteristics. The children were distinguished only by their size (Ariès, 1960, p. 34). Socially, too, children were treated just like adults. As soon as they could walk and talk, they joined adult society, playing the same games, working at the same tasks, and wearing the same kinds of clothes (Ariès, 1960, pp. 71-2; Bk. 2, Ch.1). People simply saw no difference between children and adults. It was as if the child's body and personality had been preformed into the adult pattern from the start.

Why did people hold performationist views? Ariès (1960, p. 40) speculates that for a long time people were reluctant to pay much attention to children's distinctive features because of the high rates of childhood mortality. Knowing that their children might very well die, parents were hesitant to attend to and cherish their unique qualities. According to Ausubel (1958, p. 24), preformationism also may have had to do with natural adult egocentrism. Adults tend to assume that all human life has the same form and function as their own. It takes a special open-mindedness to see the unique properties of life at different periods—an open-mindedness that is not easily acquired.

In the sciences, preformationism is most evident in the early theories of embryology. For many centuries, most scientists believed that a tiny, fully formed human, or homunculus, is implanted in the sperm or the egg at conception (see Figure 1.1). They believed the human is "preformed" at the instant of creation and only grows in size and bulk until birth. Preformationism in embryology dates back at least to the fifth century B.C. and dominated scientific thinking throughout the ages. Even as late as the eighteenth century, most scientists held preformationist views. They admitted that they had no direct

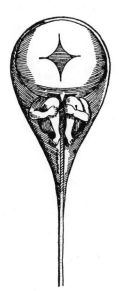

FIGURE 1.1   Drawing by Hartsoeker (1694) of a fully formed human in the sperm. (Reprinted in Needham, 1959, p. 206.)

evidence for the existence of a fully formed homunculus, but they argued that this was only because it is transparent and too small to see (Needham, 1959, pp. 34-35, 91, 213-22; Balinsky, 1970, p. 12).

In embryology, preformationism began to give way in the last half of the eighteenth century, when microscopic investigations showed that the embryo develops through a series of stages. In the realm of social thought, preformationism declined earlier, in the sixteenth century. Religious leaders and moralists began emphasizing children's special qualities. According to some, children enter the world with a God-given purity and innocence which is later corrupted. Others argued that children bear mankind's original sin. Children, they emphasized, are ignorant creatures who possess a wanton sensuality and a lack of morals (Ariès, 1960, Ch. 5).

In either case, this new interest in the distinctive nature of children brought with it a new concern for education. The proponents of childhood innocence wanted to preserve and strengthen children's innate goodness against the world's evil influence. They optimistically held that the right education would make the child into "the good magistrate," "the good priest," and "the good lord" (Ariès, 1960, p. 114). Those who viewed children more pessimistically were even more concerned about the need for education. Children's irrational and vile ways, they argued, must be forced out of them. We must impose stern discipline, including the frequent use of physical beatings. In either case—whether children were seen as innocent or sinful—they must be educated to become right-minded adults, and it is never too early to begin (Ariès, 1960, pp. 128-33).

LOCKE'S ENVIRONMENTALISM

Biographical Introduction

By the sixteenth century, then, preformationism had begun to give way to a more environmentalistic position, at least in the realm of social thought. Children, in the new view, are not born just like adults, but become the adults they do because of their upbringing and education. The writers of this period, however, were not thoroughgoing environmentalists; they still recognized certain inborn traits of childhood, such as innocence or depravity. The first clear, comprehensive statement of the environmentalist position was made by the English philosopher John Locke (1632-1704). In 1690 Locke proposed that children are neither innately good or innately bad—innately, they are nothing at all. The child's mind, Locke said, is initially a *tabula rasa*, a blank slate, and whatever the mind becomes is almost completely the result of learning and experience.

Locke's theory fit well with the liberal and democratic thinking of the Enlightenment. If children are all initially empty organisms, they are all born equal. If some people become in any way better than others, this is due to more favorable circumstances. In theory at least, it is possible to educate all people to become equal as adults.

Locke was born in the small village of Somerset, England, the son of a pious and affectionate mother and a strict father. Locke's father, who had sided with the Parliament in the civil war against King Charles I, first instilled in him a belief in democracy. Locke was educated at Oxford, where he stayed on to teach Greek and moral philosophy. There he also studied medicine. As a physician, Locke successfully treated Lord Ashley, later the Earl of Shaftesbury, became Shaftesbury's friend and personal secretary, and also tutored his grandson. His association with Shaftesbury, however, eventually proved troublesome. When Shaftesbury was imprisoned for criticizing the King, Locke was forced to flee England and find exile in Holland. While in Holland, Locke wrote a series of letters to his friend Edward Clark, offering advice on the upbringing of Clark's son. These letters inspired Locke's most important work on education, *Some Thoughts Concerning Education* (1693). After the successful Revolution of 1688, Locke returned to England and saw the publication of two other great books. The first was his *Essay Concerning Human Understanding* (1690), which established him as the father of empiricism in philosophy and learning theory in psychology. His other great book was *Two Treatises on Government* (1689), which set forth many of the central ideas in the U.S. Constitution (Sahakian and Sahakian, 1975, Ch. 1; Russell, 1945, pp. 604-5).

Locke's Views on Development

The starting point of Locke's theory was his refutation of the doctrine of innate ideas. Plato, Descartes, and others had maintained that certain ideas are innate, existing in the mind prior to experience. For example, when we see a mathematical proof, we immediately perceive its truth. Locke argued, however, that since children and idiots know nothing of mathematics or logic, these ideas cannot exist from the beginning (Locke, 1690, Vol. I, Bk. 1, Ch. 2, sec. 27). Locke suggested that we consider

> the mind to be, as we say, white paper void of all characteristics, without any *ideas*. How comes it to be furnished? . . . . Whence has it all the materials of reason and knowledge? To this I answer, in one word, from *experience*; in that all our knowledge is founded, and from that it ultimately derives itself (1690, Vol. I, Bk. 2, Ch. 1, sec. 2).

Locke admitted that individuals have special temperaments, but on the whole, the environment forms the mind (Locke, 1693, secs. 1, 32). Of particular importance is learning during infancy. At this time the mind is the most pliable, so we can mold it in any way we wish. And once we do so, its basic nature is set for the rest of life (1693, secs. 1 and 2).

How, then, does the environment shape the child's mind? First, many of our thoughts and feelings develop through *associations*. Two ideas regularly occur together, so we cannot think of one without simultaneously thinking of the other. For example, if a child has had bad experiences in a particular room, the child cannot enter it without automatically experiencing a negative feeling (Locke, 1690, Vol. I, Bk. 2, Ch. 33, sec. 15).

Much of our behavior also develops through *repetition*. When we do something over and over, such as brushing our teeth, the practice becomes a natural habit, and we feel uneasy when we have failed to perform it (Locke, 1693, sec. 66).

We also learn through *imitation*. We are prone to do what we see others do, so models influence our character. If we are frequently exposed to silly and quarrelsome people, we become silly and quarrelsome ourselves; if we are exposed to more noble minds, we too become noble (1693, sec. 67).

Finally, and most important, we learn through *rewards* and *punishments*. We engage in behavior that brings us praise, compliments, and other rewards; we refrain from actions that produce unpleasant consequences (1693, sec. 54).

These principles, Locke believed, often work together in the development of character. For example, a little girl is likely to hang up her clothes if she sees her parents hang theirs up, through imitation. After she hangs up her clothes a few times in succession, this good trait becomes a habit, and this habit becomes all the stronger if she receives some praise or compliment for it.

The above example illustrates how Locke's principles can be applied to child-rearing. Let us now look more closely at his views on child-rearing and education.

Locke's Educational Philosophy

Self-control. The main goal of education is self-control. "It seems plain to me," Locke said, "that the principle of all virtue and excellency lies in a power of denying ourselves the satisfaction of our own desires, where reason does not authorize them" (1693, sec. 38).

To instill self-discipline, we first should tend to the child's physical health. When the body is sick and weak, one has little ability to control its demands. Accordingly, Locke advised us to give children plenty of exercise so their bodies will become strong, and he suggested that children play outdoors in all seasons so they will learn to endure the hardships of all kinds of weather (1693, secs. 1-16, 33).

If children are to acquire discipline, we must be firm with them from the start. Many parents coddle their children and give in to their every whim; they think that such indulgence is all right because their children are still small. What they fail to realize is that early habits are difficult to break. Children who find that they can get whatever they want, simply by asking or crying out, never unlearn this bad habit. Thus, parents should never reward children when they desire things they do not need. Children should learn that they will get favorable results only when they ask for things that their parents consider appropriate (1693, secs. 38-40).

The best rewards and punishments. From the beginning, then, we should pay close attention to how we reinforce our children. We should only reward reasonable behavior, never behavior that is unreasonable or self-indulgent.

The use of rewards and punishments, however, is a tricky matter. Not all rewards and punishments produce desirable effects. Locke was especially opposed to the use of *physical punishment.* In the first place, its use establishes undesirable associations. For example, the child who has been repeatedly beaten or chastised in school cannot look upon books, teachers, or anything having to do with school without experiencing fear or anger. Further, physical punishment is often ineffective. The child submits while the rod is in sight, but just as soon as the child sees that no one is looking, the child does whatever he or she wants. Finally, when physical punishment does work, it usually works too well. It succeeds in "breaking the mind; and then, in the place of a disorderly young fellow, you have a low-spirited moped creature" (1693, sec. 51).

Similarly, not all kinds of rewards are desirable. Locke opposed the use of money or sweets as rewards because their use undermines the main goal of education: to curb desires and to submit to reason. When we reward with food or money, we only encourage children to find happiness in these things (1693, sec. 52).

The best rewards are praise and flattery, and the best punishment is disapproval. When children do well we should compliment them, making them feel proud; when they do poorly we should give them a cold glance, making them feel ashamed. Children are very sensitive to approval and disapproval, especially from their parents and those on whom they depend. So we can use these reactions to instill rational and virtuous behavior (1693, sec. 57).

We also can strengthen the effectiveness of our approval and disapproval by pairing these reactions with other consequences. For example, when a little boy asks politely for a piece of fruit, we give it to him, and we also compliment him on his politeness. In this way, he learns to associate approval with agreeable consequences, and thus becomes more concerned about it. Alternatively, when he breaks something he likes, we add a look of disappointment in him, so he will come to associate our disapproval with negative consequences. Through such practices, we deepen the child's concern for the opinions of others. Locke said that if you can make children "in love with the pleasure of being well thought on, you may turn them as you please, and they will be in love with all the ways of virtue" (1693, sec. 58).

Rules. Most parents set down all kinds of rules and then punish their children when they disobey them. This practice is basically useless. Children have great difficulty comprehending and remembering rules in the abstract, and they naturally resent getting punished for failing to comply with a rule which they could barely keep in mind. As an alternative to commands, Locke suggested two procedures.

First, since children learn more from example than precept, we can teach them much by exposing them to good models. Children will eagerly model their behavior after that of a virtuous person, especially when we compliment them for doing so (1693, sec. 68).

Second, Locke suggested that, instead of issuing commands, we have children practice the desired behavior. For example, instead of instructing children to bow whenever they meet a lady, it is better to give them actual practice in bowing, complimenting them each time they bow correctly. After repeated practice, they will bow as naturally as they breathe, without any thought or reflection, which is essentially foreign to them anyway (1693, sec. 66).

Children's special characteristics. Locke's discussion of the futility of teaching rules that exceed a child's understanding introduced something new into his system. Before this, he had written as if the child's mind were a lump of clay which we could mold in any way we wished. Now, however, he was saying that children have their own cognitive capacities which set limits on what we can teach. He also suggested that children have temperaments peculiar to their age, such as a liking for noise, raucous games, and gaiety, and he added that it would be foolish to try to change their natural dispositions (1693, sec. 63). Thus, Locke seemed to admit that children are not blank slates after all. As various scholars have pointed out (e.g., Russell, 1945, p. 606; Kessen, 1965, p. 59, 72), Locke was not above a certain amount of inconsistency. If he had insights that contradicted his basic environmentalism, the inconsistency didn't trouble him.

In one interesting section (1693, secs. 118-19), Locke wrote about children's innate curiosity in a way which makes one wonder about the rest of his thesis. Children, he said, learn for the sake of learning; their minds seek knowledge like the eye seeks light. If we simply listen to their questions and answer them directly, their minds will expand beyond what we would have imagined possible. But if the child's curiosity is so powerful, why do we need to use rewards and punishments for learning? Perhaps they are necessary in the development of moral behavior, but it may be that children will develop their intellectual powers through their intrinsic curiosity alone. Locke, however, never entertained this possibility, and in the end he reverted to his main thesis. When children reason clearly, we should compliment and flatter them. In this way, we teach them to reason (1693, sec. 119).

Evaluation

As a psychologist, Locke was far ahead of his time. His principles of learning—the principles of association, repetition, modeling, and rewards and punishments—all have become cornerstones of one or another version of modern learning theory. His ideas on education, furthermore, are pretty much those of the contemporary educator. Most teachers use rewards and punishments, such as praise, grades, and criticism, to motivate children to learn. Most enlightened teachers are also aware of the influence of models, of the need for repetition, and are reluctant to use physical punishment.

Most modern educators even share Locke's inconsistencies. Although they believe that it is necessary to shape or mold the child through rewards and punishments, they also recognize that such social influences are not all-powerful. They are sensitive to the child's readiness to learn different things, and they recognize that children learn best when they are spontaneously curious about a particular subject. Nevertheless, like Locke, they are not prepared to rely too heavily on children's intrinsic motivation to learn on their own. Teachers believe that it is up to them, the adults, to teach children the right things. They do not really believe that children would learn what they should without external inducements such as praise, grades, and the hope of advancing from Reading Group 2 to Reading Group 1. In general, they share Locke's view that education is essentially a socialization process. The child learns to gain our approval, and in this way the child learns what he or she needs to know to become a useful and virtuous member of society.

ROUSSEAU'S ROMANTIC NATURALISM[1]

Biographical Introduction

We have now reviewed two early conceptions of development. We have discussed the preformationist view, which considered the child as a miniature adult. We have also discussed the views of Locke, who argued that children are like empty containers which are filled in by adult teachings.

The true developmentalist position is different again. Its first forceful expression is found in the work of Jean Jacques Rousseau (1712-1778). Rousseau agreed with Locke that children are different from adults, but he made the point more positively. They are not empty containers or blank slates but have their own modes of feeling and thinking. This is because they grow accord-

---

[1] This heading is suggested by Muuss (1975, p. 27).

11

ing to Nature's plan, which urges them to develop different capacities and modalities at different stages.

Rousseau believed that it is vital for us to give Nature the chance to guide the child's growth. Unlike Locke, he had no faith in the powers of the environment, especially the social environment, to form a healthy individual. Well-socialized adults, he felt, are far too dependent on the opinions of others. They have forgotten how to see with their own eyes and to think with their own minds; they only see and think what society expects them to. So, instead of rushing in to teach children to think in the "correct" ways, we should allow them to perfect their own capacities and to learn in their own ways, as Nature intends. Then they will learn to trust their own powers of judgment.

Rousseau's beliefs, especially his faith in Nature as opposed to societal influences, sparked the Romantic Movement in the history of ideas. At the same time, his belief in a natural ground plan for healthy growth ushered in the developmental tradition in psychology.

Rousseau's revolt against society grew out of his personal life. He was born in Geneva, the son of a watchmaker and a beautiful, sentimental mother who died giving birth to him. For the first eight years of his life, he was raised by his father and an aunt. He said that his father was devoted to him, but he added that his father never let him forget that he had caused his mother's death (Rousseau, 1788, p. 5). His aunt also was kind, but she refused to let him play in the street with the other children. Rousseau therefore spent most of his time reading, and by the age of seven he had read every novel in his mother's library.

When Rousseau was eight, his father got into a bitter dispute and had to flee Geneva to avoid prison. For the next eight years, Rousseau was shuttled through several homes. He rarely got along with his masters, who often humiliated him, intensifying his already timid and self-conscious nature. He told, for example, of wanting to buy some pastry but of being afraid to enter the shop because he imagined that acquaintances would spot him and laugh at him (1788, p. 36). His main relief came from fantasies, in which he imagined himself in the heroic situations he had read about. He also engaged in a good deal of stealing and cheating.

When Rousseau was 16, he began the life of a vagabond. He travelled about, trying to earn what money he could, but was never successful. His main talent, he found, was winning the favors of older women. He was not exactly a Don Juan—he was very timid when it came to sex—but he did get several ladies to take care of him.

At the age of 33, Rousseau took up with an illiterate servant girl named Thérèse, with whom he spent the rest of his life. She gave birth to five children, but Rousseau placed them all in a state foundling home. He said that he later realized that this action was wrong, but at the time he did not have the money to raise them, and he felt that if he did they would wind up living a life as miserable as his own (1788, p. 367).

Rousseau's literary career began at the age of 37, when he entered an essay contest which asked whether the arts and sciences had contributed to the betterment of morals. Rousseau argued in the negative and won the prize (Rousseau, 1750). During the next several years, he wrote several essays and books, the most important of which are *The Social Contract* (1762a) and *Emile* (1762b). *The Social Contract* opens with the famous line, "Man is born free, and everywhere he is in chains." That is, humans are naturally good and could live happily according to their spontaneous passions, but they are enslaved by social forces. This book describes a better society. *Emile* is Rousseau's main book on child development and education. It is titled after the fictitious boy whom Rousseau proposed to tutor according to Nature's plan for healthy development.

In the course of his writings, Rousseau challenged the divine rule of kings and various religious orthodoxies and argued for a form of democracy. As a result, Geneva burned his books and France banished him from the country. He spent many of his last years in exile, paranoid and miserable. When Rousseau died, he was buried in the French countryside, where his ashes remained until after the French Revolution, which his writings had helped inspire. His remains were then triumphantly removed to Paris and placed in the Pantheon.

Many people have found Rousseau so deficient as a man that they have refused to take his ideas seriously, especially on education. How can a man who abandoned his own children to an orphanage have the audacity to prescribe the right upbringing for others? However, it sometimes takes one who has lived on the outside of the conventional social order to create a radical vision. Rousseau said that he was "thrown, in spite of myself, into the great world, without possessing its manners, and unable to acquire or conform to them . . ." (Rousseau, 1788, p. 379). He felt that his only legitimate response was to rail against society and to seek, in its place, a different vision of how life might unfold. He tried to show how the healthiest development might come not from society's influence, but from Nature. In so doing, Rousseau became the father of developmental psychology.

Rousseau's Theory of Development

Childhood has a special place in the sequence of human life, yet we know nothing about it. This is because we are so exclusively concerned with the child's future—with the things the child will need to know to fit into adult society. Even "the wisest writers devote themselves to what a man ought to know, without asking themselves what a child is capable of learning. They are always looking for the man in the child, without considering what he is before he becomes a man" (Rousseau, 1762b, p. 1).

When we take the time simply to observe children, we find that they are very different from us. "Childhood has its own ways of seeing, thinking, and feeling . . ." (1762b, p. 58). This is according to Nature's design. Nature is like a hidden tutor who prompts the child to develop different capacities at different stages of growth (1762b, p. 181). Her product might not be an individual well-trained to fit into a social niche, but a strong, complete person. If we wish to aid Nature in this process, we must first learn all we can about the stages of development. Rousseau believed that there are four main stages.

Stage 1: Infancy (birth to about two years). Infants experience the world directly through the senses. They know nothing of ideas or reason; they simply experience pleasure and pain (1762b, p. 29). Nevertheless, babies are active and curious and learn a great deal. They constantly try to touch everything they can, and by doing so they learn about heat, cold, hardness, softness, and other qualities of objects (1762b, p. 31). Infants also begin to acquire language, which they do almost entirely on their own. In a sense, they develop a grammar that is more perfect than ours; they employ grammatical rules without all the exceptions that plague adult speech. Pedantically, we correct their mistakes, even though children will always correct themselves in time (1762b, p. 37).

Stage 2: Childhood (about two to 12 years). This stage begins when children gain a new independence; they can now walk, talk, feed themselves, and run about. They develop these abilities, too, on their own (1762b, p. 42).

During this stage, children possess a kind of reason, but it is not the kind that deals with remote events or abstractions. Rather, it is an intuitive reason that is directly tied to body movement and the senses. For example, when a girl accurately throws a ball she demonstrates an intuitive knowledge of velocity and distance. Or when a boy digs with a stick, he reveals an intuitive knowledge of leverage. However, thinking is still extremely concrete. Rousseau told, for example, about a boy who had made a globe, with all the countries, towns, and rivers. When he was asked, "What is the world?" he replied, "A piece of cardboard" (1762b, p. 74).

Stage 3: Late Childhood (about 12 to 15 years). This third stage is a transitional one between childhood and adolescence. During this period, children gain an enormous amount of physical strength; they can plough, push carts, hoe, and do the work of adults (1762b, p. 128). They also make substantial progress in the cognitive sphere and can, for example, do relatively advanced problems in geometry and science. Still, they are not yet disposed to think about purely theoretical and verbal matters. Instead, they can best exercise their cognitive functions through concrete and useful tasks, such as farming, carpentry, and map-making.

During the first three stages, children are by nature *presocial*. That is, they are primarily concerned with what is necessary and useful to themselves and have little interest in social relationships. They enjoy working with physical things and learning from nature; the world of books and society is foreign to them. Even as late as the third stage, between 12 and 15 years, the model for the child's life should be Robinson Crusoe, a man who lived alone on an island and who became self-sufficient by dealing effectively with the physical environment (1762b, p. 147).

Stage 4: Adolescence. Children become distinctly social beings only at the fourth stage, which begins with puberty. Rousseau said that puberty begins at age 15, somewhat later than we would date it today. At this time, the child undergoes a second birth. The body changes and the passions well up from within. "A change of temper, frequent outbreaks of anger, a perpetual stirring of the mind, make the child almost ungovernable" (1762b, p. 172). The young person, who is neither child nor adult, begins to blush in the presence of the opposite sex, for he or she is dimly aware of sexual feelings. At this point, the youngster is no longer self-sufficient. The adolescent is attracted to and needs others.

The adolescent also develops cognitively. He or she can now deal with abstract concepts and takes an interest in theoretical matters in science and morals.

These, then, are Rousseau's four stages, which he believed unfold in an invariant sequence according to Nature's plan. Rousseau also proposed that these stages *recapitulate* the general evolution of the human species. Infants are similar to the earliest "primitives," who dealt with the world directly through their senses and were concerned only with pleasure and pain. The next two stages of childhood parallel the "savage" era, when people learned to build huts, make tools, fish, trap, and utilize other skills. People formed loose associations with others, but they still were largely self-sufficient.

Adolescence, finally, parallels the beginning of true social life. Historically, social existence began with the division of labor. As work became specialized, people could no longer produce all they needed by themselves. Thus, they had to rely on others. As they became increasingly immersed in society, they became the slaves of conventions and social approval. Even savages, to be sure, were somewhat concerned with the opinions of others, but this concern deepened as people became embedded in social life. As a result, modern individuals no longer think for themselves. "The savage," Rousseau said, "lives within himself; the sociable man, always outside himself, knows how to live only in the opinion of others" (Rousseau, 1754, p. 179).

Rousseau's Educational Method

Rousseau thought we were most fulfilled as savages, but he realized that those days are gone forever. Still, we do not need to become the weak conformists that we presently are. Nature will still guide children's development along the road to independence. Under her urging, children will spontaneously perfect their capacities and powers of discrimination by dealing with physical things, without adult teaching. So, if one followed Nature's guidance, it should be possible to bring the child to adolescence with an independent mind. Then, when the young person did enter the social world, he or she could cope effectively with it.

Rousseau told how this would happen in the case of Emile, his imaginary pupil.

Emile's education. Rousseau would have a basic faith in Emile's capacity to learn much on his own, from Nature's inner promptings. For example, as an infant Emile would have a strong urge to explore the world through his senses. Accordingly, Rousseau would remove all harmful objects from the house and let Emile explore it. If Emile wished to inspect an object, Rousseau would bring it to him. No adult guidance would be necessary (Rousseau, 1762b, p. 31, 35).

At the same time, Rousseau would not permit Emile to rule over him. He would bring Emile an object when Emile had a genuine need to learn about it, but never when Emile simply had a capricious desire to have his tutor do his bidding (1762b, p. 52).

Emile also would learn to walk and talk on his own. Rousseau would never push or correct his pupil. Such practices only make children timid and anxious. They begin looking to others for correction and thereby lose their independence (1762b, pp. 39-40).

As Emile moved into the second stage, that of childhood, he would have an urge to run, jump, shout, and play. Rousseau would never check these activities, for Emile would be following Nature's inner prompting to develop his body through vigorous exercise. Rousseau would not, like many adults, always be saying, "Come here, go there, stop, do this, don't do that" (1762b, p. 82). For Emile would then turn to his tutor for guidance and "his own mind would become useless" (1762b, p. 82).

Rousseau would present various lessons, but only those that fit Emile's age. Since children at this stage are developing their senses, Rousseau would suggest games such as finding one's way in a completely dark room, thus developing the sense of touch (1762b, p. 98). Since children do anything that keeps them moving freely, he would take advantage of this impulse to help Emile learn to judge heights, lengths, and distances. He would point to a cherry tree and ask Emile to select a ladder of the proper height. Or he would suggest that they cross a river, and ask Emile which plank would extend across the banks (1762b, p. 105).

In all such lessons, Emile would be able to judge his successes by himself. Emile could see for himself, for example, if he had chosen a plank that was large enough to extend across the river. He could make this judgment because the lesson corresponds to his current capacities. It requires only the use of his senses. There is nothing in the lesson that is beyond his grasp, nothing that would force him to turn to his tutor for help (1762b, p. 141).

During the third stage, that of late childhood, Emile's maturing cognitive powers would enable him to learn mathematics and science, but he would reason effectively in these spheres only in connection with concrete activities. Accordingly, Rousseau would encourage him to think about mathematical problems that naturally emerged in the course of activities such as farming and carpentry. Rousseau would provide only minimal guidance and, again, he would never correct Emile's mistakes. His goal would not be to teach Emile the right answers but to help him learn to solve problems on his own.

> Let him know nothing because you have told him, but because he has learned it for himself. Let him not be taught science, let him discover it. If ever you substitute authority for reason he will cease to reason; he will be a mere plaything of other people's thoughts (1762b, p. 131).

Only at adolescence would Emile begin reading many books and receive his introduction into the larger social world. By this time he would have developed an independent nature, and with his new capacities for theoretical reasoning, he could judge society at its true worth (1762b, p. 183).

Comparison with the usual practices. Rousseau, then, would encourage Emile to perfect his capacities at each stage, according to Nature's own schedule, and he would never present anything that Emile could not judge for himself. His method would differ radically from that of most educators.

Most teachers are not content to treat children as children, with their own needs and ways of learning. Instead, they try to instill adult knowledge as quickly as possible. As a result, they present many lessons that exceed the child's understanding. For example, they give lessons in history, geography, and mathematics that have nothing to do with the child's direct experience and which assume a capacity for reasoning which the child lacks. As children struggle with such lessons, they find learning a miserable experience. And not only this. Because they cannot fully comprehend what the adult is saying, they are forced to take things on faith, to accept answers simply because the adult has explained them to be true. They have no recourse but to ask their parents or teachers, "Did I get the right answer here?" "Is this right?" They thereby learn to depend on others and cease to think for themselves.

When children are asked to learn things that exceed their grasp, they become lazy and unmotivated. To motivate them, teachers use threats, bribes, disapproval, flattery, and other social reinforcers. They try to get children to learn in order to win the adult's approval (1762b, p. 54). Such procedures only reinforce the child's dependency on the approval of others.

Rousseau said that his own method, in contrast, would be "merely negative" (1762b, p. 57). That is, he would exercise Emile's body and senses but keep his mind idle as long as possible. He would shield Emile from all opinions until his capacity for reasoning had developed, at which point he could judge them for himself. At the age of 12 or 15 years, Emile would appear ignorant in the conventional sense. He would know nothing of society or morals and have no precocious knowledge to display. He would be nothing but a rough, happy boy. But he would have learned to judge everything according to his own experience. He therefore would be capable of real thinking (1762b, p. 127, 170).

Rousseau anticipated the impatience others would have with his advice. It seemed as if he were failing to prepare the child for the future. How could we be certain that the child would know what was necessary when the time came? Rousseau's reply was that children cannot genuinely learn something until they are ready to grasp it (1762b, p. 141). He also observed that societies change so rapidly that it is impossible to predict what knowledge will be useful anyway (1762b, p. 157). What is important is not that children acquire specific information, but that they learn to think for themselves. And they cannot possibly do this if they are forced to accept beliefs that are beyond their understanding.

Evaluation

Rousseau introduced several key ideas into developmental theory. First, he proposed that development proceeds according to an inner, biological timetable. For the first time, we have a picture of development unfolding fairly independently from environmental influences. Children are no longer simply shaped by external forces, such as adult teachings and social reinforcements. They grow and learn largely on their own, according to Nature's plan. Today we would call this plan biological maturation.

Second, Rousseau suggested that development unfolds in a series of stages, periods during which children experience the world in different ways. Children differ from adults not because they are blank slates which will gradually take on adult teachings; rather, at each stage, the child's patterns of thought and behavior have their own unique characteristics.

Third, Rousseau proposed a new philosophy of education, one which we would today call "child-centered." He said, "Treat your scholar according to his age" (1762, p. 55), by which he meant we should fit our lessons to the child's particular stage. In this way, children will be able to judge matters according to their own experience and powers of understanding.

All three of these ideas have become central tenets of many developmental theories; this is shown in the following chapters. At the same time, though, many developmental theorists would disagree with parts of Rousseau's theory. Many would argue, in particular, that the child is not nearly as asocial as Rousseau suggested. For example, modern ethologists point out how babies become strongly attached to their caretakers. This attachment, they say, is genetically governed; it has evolved because proximity to parents has enhanced babies' chances for survival (See Chapter 3 in this book). Actually, Rousseau was aware of such attachments (1762b, p. 174), but he conveniently ignored them when outlining his overall theory. He wanted children to learn to reason on their own, apart from society's corrupting influences, and he therefore declared that Nature intends for them to live apart from the social world, even if he knew better.

When Rousseau argued that we should protect children from society, he had particular concerns in mind. He saw adults teaching children social manners and beliefs before children have the ability to judge them according to their own powers of reasoning. In this process, adults make children the slaves of social conventions.

Today, however, there are developmentalists (e.g., Kohlberg, Chapter 6 in this book) who prize independent thinking as much as Rousseau did, but who believe, nevertheless, that children can make their way through the social world. They believe that children will form social and moral theories on their own, fairly independent of adult teachings. Furthermore, if children think long and hard about social problems, they will reach stages that transcend conventional modes of social thought. Thus, it may be that children can live in the social world without being undone by it.

All the same, it was Rousseau who introduced the crucial question into modern developmental and humanistic thinking: Can inner growth lead to ways of experiencing and feeling that can stand up to the crushing pressure of social conformity?

Gesell's Maturational Theory

2

BIOGRAPHICAL INTRODUCTION

When Rousseau said that development unfolds according to a set schedule, he was only guessing. In his day, there was little scientific evidence to support such a view, and none that he was aware of. The first evidence for sequential growth came from eighteenth-century studies in embryology. Scientists, using new and powerful microscopes, saw that there was no fully formed organism in the sperm or egg, but that the embryo develops in a series of steps (epigenetically), and always in the same sequence. Since they found this to be the case in the embryo, there was reason to look for sequential unfolding of structures after birth as well.

The person who most thoroughly applied the embryological model to the study of child development was Arnold Gesell (1880-1961). Gesell grew up in Alma, Wisconsin, a small town on the bank of the upper Mississippi River. In an autobiographical account, he described an almost idyllic Midwestern childhood, in which "hills, valley, water and climate concurred to make the seasons distinct and intense in my home town. Each season had its own challenges and keen pleasures, accentuated by the everchanging, yet enduring river" (Gesell, 1952a, p. 124). Gesell used similar language to describe the beauty he saw in the growth process, with "its seasons and sequences" (Gesell and Ilg, 1943, p. 57). This is not to say, however, that Gesell was merely a gushing romantic. He studied children's development with painstaking observation. To increase his knowledge of the underlying physiological processes, he went to medical school at the age of 30, even though he already had a Ph.D. and had been working successfully as a psychologist. In his 50 years at the Yale Clinic of Child Development, he and his colleagues engaged in incredibly extensive and detailed studies of the neuromotor development of babies and children. They developed behavior norms which are so complete that they still serve as a primary source of information for pediatricians and psychologists. Gesell also developed one of the first tests of infant intelligence (Gesell and Amatruda, 1941) and was one of the first researchers to make extensive use of film observations.

Gesell also wrote on child-rearing, advocating a child-centered approach. He was the best known "baby doctor" in the early 1940s, until Spock published his famous book (1945). Nevertheless, Spock was partly influenced by Gesell.

PRINCIPLES OF DEVELOPMENT

The Concept of Maturation

Human life begins as a single tiny cell. This cell attaches itself to the uterine wall and rapidly divides and multiplies. As the embryo becomes increasingly differentiated, the cells gather to form different parts of the organism. They always do so in an orderly sequence. For example, the heart is always the first organ to develop and function. Soon afterward, the cells rapidly begin to form the central nervous system—the brain and spinal cord. The development of the brain and head, in turn, begins before other parts, such as the arms and legs (see Figure 2.1). For some time, the head is quite large in comparison to other parts (Mussen *et al.*, 1974, pp. 96-103).

A milestone is reached eight weeks after conception: neuromotor activity begins. The head and trunk make slight movements. At this point the embryo is called a fetus. The activation of the neuromotor system also occurs in set sequences (Gesell, 1945, Ch. 6).

Prenatal development is largely controlled by the genes, chemicals which are contained within the nucleus of each cell (see Figure 2.2). The genes initiate and direct the sequential unfolding of organic forms and action patterns. However, the precise role of the genes is still mysterious, and they do not act alone. They appear to receive, for example, some cues from the cytoplasm, the cellular substance outside the nucleus (Gesell, 1945, p. 22; Balinsky, 1970, pp. 553-56). Despite the many unknowns, Gesell gave a general name to the mechanism by

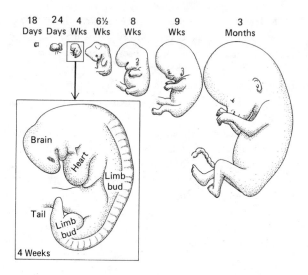

| 18 | 24 | 4 | 6½ | 8 | 9 | 3 |
| Days | Days | Wks | Wks | Wks | Wks | Months |

Brain

Heart

Limb bud

Tail

Limb bud

4 Weeks

FIGURE 2.1 The growing embryo. *Adapted from* Vincent, E. L., and Martin, P. C. *Human Psychological Development.* Copyright © 1961, The Ronald Press, p. 94. By permission of John Wiley and Sons, Inc.

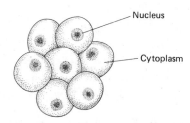

Nucleus

Cytoplasm

FIGURE 2.2 A group of cells. The nucleus contains chromosomes, made up of genes. *Adapted from* Munn, N. L. *The Growth of Human Behavior,* 3rd ed. Boston: Houghton Mifflin Co., 1974, p. 40. By permission.

which genes direct the developmental process. He called this mechanism *maturation* (1952b, p. xi, 64).

Maturation is distinguished from the role of the environment. In prenatal development, this means that maturation is distinguished from aspects of the internal environment, such as the embryo's temperature and the oxygen it receives from its mother. These environmental factors support proper growth, but they play no direct role in the sequential unfolding of structures and action patterns. This is the work of the maturational mechanism.

After birth, maturation continues to direct development in specific sequences. For example, just as the head develops early in the embryo, it also takes the lead in early postnatal development. Babies first have control over their lips and tongue, then gain control over their eye movements, followed by control of the neck, shoulders, arms, hands, fingers, trunk, legs, and feet. In both prenatal and postnatal development, there is a head-to-foot (cephalocaudal) trend (Gesell, 1946, p. 339). As babies grow, they continue to develop their capacities in a specific order. For example, they always sit up before they stand and stand before they walk. These capacities develop with the growth of the nervous system, which itself is mediated by the genes.

Children vary, of course, in their rates of development. They do not all stand up and walk at the same age. Nevertheless, they all proceed through the same sequences. Furthermore, even individual differences in growth rates, in Gesell's view, are largely controlled by internal genetic mechanisms (1945, p. 161).

In postnatal development, maturation is clearly distinguished from the influences of the external environment. In particular, maturation is contrasted with the effects of teaching or practice. When a skill is governed by maturation, there is little point to trying to teach it ahead of schedule. Children will sit up, walk, and talk when they are ready, when their nervous systems have sufficiently matured. At the right moment, they will simply begin to master a task, from their own inner urges. Until then, teaching will be of little value.

Evidence for the maturational position has come from studies with identical twins. For example, Gesell and Thompson (1929) gave one twin practice at such activities as stair-climbing, block-building, vocabulary, and manual coordination. This twin did show some skill superior to that of the other, but the untrained twin soon caught up, with much less practice. And he did so at about the age at which we would expect him to perform the various tasks. Apparently, then, there is an inner timetable which determines the readiness to do things, and the benefits of early training are relatively temporary. The question of early stimulation is controversial, but when developments are strongly governed by intrinsic maturational factors, teaching them ahead of schedule seems to have only small effects (Munn, 1974, pp. 82-94).

This does not mean that Gesell considered the environment unimportant. He pointed out (e.g., 1952b, p. xi) that specific environmental conditions are necessary for normal growth, as we can observe when there is damage to the organism. For example, if the growing embryo suffers a sharp drop in oxygen, the organ that is undergoing the most rapid growth at this time will be severely impaired. (This period of rapid growth of an organ is called its "critical period.")

Similarly, normal postnatal development requires certain environmental conditions. For example, institutionalized children whose environment is extremely impoverished in stimulation and care do not develop well (Gesell and Ilg, 1943, p. 34). Their motor or social development may lag behind that of normal children, at least until added stimulation is provided (Ainsworth, 1973). The child needs, then, a favorable environment to insure the favorable realization of his or her potentials. Nevertheless, while "environmental factors support, inflect, and modify" growth, "they do not generate the basic progressions of development" (Gesell and Ilg, 1943, p. 20). These come from within.

Gesell's most important research was on early motor development, but he believed that maturation governs every aspect of growth. He said, for example,

> [The child's] nervous system matures by stages and natural sequences. He sits before he stands; he babbles before he talks; he fabricates before he tells the truth; he draws a circle before he draws a square; he is selfish before he is altruistic; he is dependent on others before he achieves dependence on self. All his capacities, including his morals, are subject to the laws of growth (Gesell and Ilg, 1943, p. 11).

The Study of Patterns

Gesell said that when we study growth we should not just measure things in quantitative form but should examine patterns. A pattern may be anything that has a definite shape or form—for example, an eyeblink. But what is most important is the patterning process, the process by which actions become organized (Gesell and Ilg, 1943, pp. 16-17).

We find good illustrations of the patterning process in the case of babies' vision. At birth, for example, babies' eyes are apt to rove around aimlessly, but after a few days or even hours babies are able to stop their eyes and look at objects for brief periods. They can stop their eyes and stare "at will" because a new patterned connection has been made between the nerve impulses in the brain and the tiny muscles which move the eyes (Gesell and Ilg, 1943, pp. 17-18).

By one month, babies can usually regard a ring that is dangled before them and then follow it through an arc of about 90°. This ability implies a new organization—that between the eye muscles and the grosser muscles which rotate the head (1943, p. 19).

Patterning continues to widen when babies organize their eye movements with their hand movements, when they look at what they hold. By four months, babies usually can hold a rattle and look at it at the same time. "This is a significant growth gain. It means that eyes and hands are doing team work, coming into more effective coordination. Mental growth cannot be measured in inches and pounds. So it is appraised by patterns" (Gesell and Ilg, 1943, p. 19).

Still, hand-eye coordination is by no means complete at four months. For some time, the eyes will be in the lead. At four months, for example, babies can often "pick up" a one-inch cube or even a small candy pellet with their eyes; that is, they can focus intently on the cube or pellet and consider it from slightly different angles. But they cannot yet grasp it with their hands. Babies may be seen looking at the cube and then looking at their hands, as if they have an idea of grasping the cube, but they simply cannot do it yet. The nervous system has not yet sufficiently grown. It is not until six months that babies usually are able to pick up the cube with a crude palmar grasp, and not until 10 months that they can pick up the small candy pellet with a pincer grasp, with opposing thumb and index finger (see Figure 2.3). Thus, hand-eye coordination develops slowly—it becomes gradually more organized and comes to include more differentiated or refined movements.

Other Principles of Development

Gesell's observations suggested several other principles of growth. We will consider three: reciprocal interweaving, functional asymmetry, and self-regulation.

Reciprocal interweaving. Human beings are built on a bilateral basis: we have two hemispheres of the brain, two eyes, two hands, two legs, and so on. "Reciprocal interweaving" refers to the process by which the two sides gradually reach an effective organization. For example, in the development of handedness, the baby first uses one hand, then both together, then prefers the other hand, then both together again, and so on until he or she ultimately reaches one-handed dominance. The back-and-forth quality of the preferences suggests the metaphor of weaving; hence the term "reciprocal interweaving." Gesell believed that the principle of reciprocal interweaving applies to a wide range of behaviors, including the patterning of prone behavior (crawling and walking), visual behavior, and children's drawings (Gesell, 1946, pp. 342-49).

Functional asymmetry. Gesell observed that humans, at the highest stages of development, usually function most effectively when they face the world not from a position of frontal symmetry but from an angle. He called this tendency to attack from the side the "principle of functional asymmetry," and he included under this principle the dominance of one hand, one foot, one eye, and so forth.

(a) Four months: sees but cannot contact.

(b) Six months: palmar grasp.

(c) Ten months: pincer grasp.

FIGURE 2.3 Developments in hand-eye coordination. *Adapted from Vincent, E. L., and Martin, P. C. Human Psychological Development.* Copyright © 1961, The Ronald Press, p. 130. By permission of John Wiley and Sons, Inc.

The infant's asymmetric tendency is seen in the *tonic neck reflex,* a reflex which Gesell discovered in humans. Gesell noted that babies prefer to lie with their heads turned to one side, and when they do so they automatically assume the tonic neck reflex posture: They extend the arm on the side to which the head is turned (as if looking at the hand) and flex the other arm behind the head. The tonic neck reflex posture looks very much like the basic stance of a fencer (see Figure 2.4). Gesell offered several speculations regarding the adaptive value of this reflex; perhaps the most interesting speculation is that this position facilitates hand-eye coordination. The tonic neck reflex is dominant during the first three months after birth and then eclipses, with new developments in the nervous system (Gesell, 1946, pp. 349-54).

Self-regulation. In all of growth, there is a considerable amount of self-regulation. The self-regulatory principle is at work when children resist our efforts to teach them too much too soon. If they were too dependent on varying environmental pressures to learn, the integrity of the organism would be endangered (Gesell, 1952b, p. 69).

The principle of self-regulated growth also manifests itself in other ways. In infancy, self-regulation underlies babies' ability to regulate their own cycles of feeding, sleep, and wakefulness. Gesell showed that when parents permit babies to determine when they need to nurse and sleep, the babies gradually require fewer feedings per day and stay awake for longer periods during the daytime. There are many fluctuations, but babies gradually work out stable patterns (Gesell, 1946, pp. 358-64). In the realm of feeding, in addition, there is some evidence that babies, when given a free choice, will select a balanced diet over time (Davis, 1939). There seems to be an inner "wisdom of the body" (Gesell and Ilg, 1943, p. 48).

Oddly enough, the mechanism of self-regulation also underlies the uneven, fluctuating nature of growth. Children thrust forward into new areas and then partially retreat, consolidating their progress before moving forward again. For example, as babies make progress in crawling and walking, they partially revert

FIGURE 2.4   The tonic neck reflex in two newborns. (Courtesy of Ellen Crain and the Jacobi Intensive Care Nursery, Bronx, New York.)

to earlier stages three or four times. Similarly, as they gradually reduce their amounts of daytime sleep, they repeatedly fluctuate between more and less, as if something inside tells them not to advance too rapidly (Gesell, 1946, pp. 358-64).

Fluctuations characterize growth in every sphere, including the child's emotional development. For example, two-and-a-half-year-olds are struggling with new developments and are in a phase of disequilibrium. They negativistically want to make all their own choices, yet have trouble doing so. Three-year-olds have resolved these problems, at least for the time being; they are more self-composed. Four-year-olds thrust forward once again; they are expansive, argumentative, and quarrelsome (Gesell and Ilg, 1946, pp. 54-56). All development proceeds through cycles such as these, through periods of stability and instability. Tensions arise as children try out new ways of acting, but children never go too far before restabilizing. Nature prevents the child from overgrowing.

Individuality

We have now reviewed some, though by no means all, of Gesell's ideas about growth. There is, however, one general issue that needs to be discussed. This is the problem of individuality. Gesell strongly believed in the uniqueness of each child. Unfortunately, however, his position was obscured by the way in which he summarized his findings. For example, he wrote about the child at two, two and a half, three, and so on as if we could expect all children at each age to behave in exactly the same way. He did warn that he was using age norms only as short-cut devices (Gesell and Ilg, 1943, pp. 60-61), but he never indicated the actual amount of individual variation that does occur at each age, so individuality did often become lost in his writings.

Gesell's actual position, as mentioned, was that normal children all go through the same *sequences,* but they vary in their *rates* of growth. He also suggested that growth rates might be related to differences in temperament and personality. In an interesting discussion (Gesell and Ilg, 1943, pp. 44-45), he presented three hypothetical children—one who grows slowly, one who grows rapidly, and one who grows irregularly—and suggested how each growth style might show up in a variety of personal dispositions. Child A, who grows slowly, might, for example, be generally slow and cautious, able to wait, even-tempered, and generally wise about life's problems. Child B, who develops rapidly, might be quick-reacting, blithe and happy, up and at it, and generally bright and clever. Child C, who develops irregularly, might sometimes be overcautious and sometimes undercautious, often moody, have trouble waiting, and show flashes of brilliance. Gesell believed that each individual temperament and growth style makes different demands on the culture, and that the culture should try to adjust to each child's uniqueness.

PHILOSOPHY OF CHILD-REARING

Gesell believed that child-rearing should begin with a recognition of the implicit wisdom of maturational laws. Babies enter the world with an inborn schedule which is the product of at least three million years of biological evolution; they are preeminently "wise" about their needs, and what they are ready and not ready to do. Thus parents should not try to force their children into any preconceived pattern, but should take their cues from the children themselves.

On the topic of feeding, for example, Gesell strongly advocated demand feeding—feeding when the baby indicates a readiness—as opposed to feeding by any predetermined schedule. He wrote,

> There are two kinds of time,—organic time and clock time. The former is based on the wisdom of the body, the latter on astronomical science and cultural conventions. A self-demand schedule takes its departure from organic time. The infant is fed when he is hungry; he is allowed to sleep when he is sleepy; he is not roused to be fed; he is "changed" if he fusses on being wet; he is granted some social play when he craves it. He is not made to live by the clock on the wall, but rather by the internal clock of his fluctuating needs (Gesell and Ilg, 1943, p. 51).

As parents suspend their ideas about what the baby "ought" to be doing—and follow the baby's signals and cues—they begin to appreciate the baby's inherent capacity for self-regulated growth. They see how the baby regulates his or her own cycles of feeding, sleep, and wakefulness. A little later, they see how the baby learns to sit up, creep, and crawl on his or her own, without pushing and prodding. Parents begin to trust the child and the growth process.

Gesell emphasized that the first year is the best time for learning to respect the child's individuality (Gesell and Ilg, 1943, p. 57). Parents who are alertly responsive to the child's needs during infancy will naturally be sensitive to the child's unique interests and capacities later on. They will be less inclined to impose their own expectations and ambitions on the child and more inclined to give the child's individuality a chance to grow and find itself.

Gesell said that parents need, besides an intuitive sensitivity to the child, some theoretical knowledge of the trends and sequences of development. In particular, they need to realize that development fluctuates between periods of stability and instability. This knowledge fosters patience and understanding. For example, the parent who recognizes that the two-and-a-half-year-old's obstinacy is a natural phase of growth will not feel an urgent need to stamp out this behavior before it is too late. Instead, the parent will be able to deal with the child more flexibly, and perhaps even enjoy the child who so intently tries to establish his or her independence (1943, p. 179, 296).

One of Gesell's followers, Louise Bates Ames (1971), offers parents the following advice:

1. Give up the notion that how your child turns out is all up to you and there isn't a minute to waste.

2. Try to appreciate the wonder of growth. Observe and relish the fact that every week and every month brings new developments . . .

3. Respect his immaturity. Anticipate the fact that he will, in all likelihood, need to creep before he walks, express himself with single words before he talks in sentences, say "No" before he says "Yes".

4. Try to avoid thinking always in terms of what comes next. Enjoy, and let your child enjoy each stage he reaches before he travels on . . . (p. 108, 125).

So far, Gesell's philosophy sounds like one of extreme indulgence and permissiveness. One might ask, "Doesn't this attitude lead to spoiling?" "Won't children start becoming bossy, always wanting their own way?"

Gesell's answer was that of course children must learn to control their impulses and get along with the demands of their culture. However, he argued that children best learn to do this when we pay attention to their own maturing ability to tolerate controls. For example, with respect to feedings, the baby at first should not be made to wait too long. "The most vital cravings of the infant have to do with food and sleep. These cravings have an individual, organic nature. They cannot be transformed or transgressed" (1943, p. 56). But a little later—by four months or so—the baby's gastrointestinal tract no longer dominates life as it did before, and the baby's less intense and less frequent cries tell the parent that he or she is now able to wait for feedings.

Later on, developments in language and an increased time perspective help children delay immediate gratification. At two and a half years they do not need their juice immediately because they understand when the parent says, "Pretty soon." At three they may understand "When it's time." And by four they want to help prepare the meals themselves. The culture, then, can ease children into its fabric by gearing itself to children's own maturing readiness to tolerate controls (1943, p. 54).

Thus, Gesell believed that alert caretakers can achieve a reasonable balance between maturational forces and enculturation. However, it is clear that Gesell wanted the culture to do most of the adjusting. Enculturation, he said, is necessary, but our first goal should not be to fit the child into the social mold. This is the aim of authoritarian regimes. In democracies we prize autonomy and individuality, qualities which have their deepest roots in the biological impulse toward optimal growth (1943, p. 10).

Enculturation takes place in the school as well as in the home. Schools teach children the skills and habits they will need as adult members of society. But teachers, like parents, should not think so exclusively in terms of cultural goals that they overlook the manner in which the child grows. For example, although our culture values accurate work, teachers need to recognize that children are naturally less precise at one age than another. Vigorous, unstable six-year-olds are error prone, whereas more stable seven-year-olds readily take to drills and perfectionistic work. Accordingly, the developmentally minded teacher will not force six-year-olds to learn in a way that runs counter to their nature, but will save drills for the time when the child benefits from them (Gesell and Ilg, 1946, pp. 374-81).

At the same time, it is not enough to adjust techniques to each age or grade; for children also vary widely in their growth rates, as well as in their special talents. Accordingly, teachers need to gear their work to the individual child's state of readiness and special abilities. At present, most schools do not do this. They overemphasize uniform achievement standards, such as grade-level reading scores, thereby ignoring children's need to grow according to their own timing and to develop their unique potentials. Schools, of course, do need to teach standard cultural skills, but in a democracy their first task is to help children develop their full personalities. To do this, they must let children guide them, just as children themselves are guided by a biological ground plan for optimal growth (Gesell and Ilg, 1946, pp. 388-92).

EVALUATION

In Gesell's hands, Rousseau's idea of an inner developmental force became the guiding principle behind extensive scholarship and research. Gesell showed how the maturational mechanism, while still hidden, manifests itself in intricate developmental sequences and self-regulatory processes. Gesell indicated that there are good reasons to suppose that development follows an inner plan.

Nevertheless, most contemporary psychologists would consider Gesell's maturational position too extreme. Most psychologists acknowledge the role of maturation but nevertheless believe that teaching and learning are much more important than Gesell claimed. They believe that the environment does more than merely support inner patterning; it also structures behavior. For example, although children cannot learn to throw a ball or play a piano before attaining some level of neuromotor maturation, they also acquire such behavior patterns through teaching and reinforcement. Still, it is largely because of Gesell's work that even most ardent learning theorists take some notice of inner maturational processes.

The most frequently voiced criticisms of Gesell center on his manner of presenting age norms. As mentioned, his norms imply too much uniformity and give us no idea of how much variation to expect at any given age. Moreover, Gesell's norms were based on middle-class children in a university setting (Yale) and may not apply perfectly in other cultural contexts.

Nevertheless, Gesell's norms, particularly with respect to infant motor development, are extremely valuable. His observations were so careful and detailed that they are still useful to pediatricians, educators, and psychologists working with children. For example, when pediatricians today examine the possibility of mental retardation, they rely on data on what children can be expected to do at various ages, and this data often is based on Gesell's observations. Although there is a need to extend his research to wider populations, his norms provide some of the most useful guidelines we presently have.

Gesell also provided a coherent philosophy of child-rearing. We should not, he said, try to force children into our predetermined designs, but should follow their cues as they express basic biological forces of growth. The research findings bearing on Gesell's position have often been ambiguous (Caldwell, 1964), but some interesting recent studies do support him (Appelton *et al.*, 1975). In a particularly impressive study, Bell and Ainsworth (1972) asked what happens when parents respond promptly to their babies' cries (rather than acting on their own ideas about when it is all right for them to cry). The clear finding was that responsiveness does not lead to spoiling. On the contrary, by the age of one year these babies, in comparison to babies of less responsive parents, cried less and were more independent. They enjoyed being held, but if the mother put them down they did not cry or protest but happily ventured off into exploratory play. They might check back on the mother's presence from time to time, as is natural at this age, but they were basically quite independent. Apparently, then, when babies' signals are heeded, they become confident that they can always get help when needed and therefore can relax and venture forth on their own.

There also is some evidence, still only impressionistic and anecdotal, that things can go very wrong when Gesell's principles are excessively violated. This evidence comes from the study of schizophrenic patients, whose childhood experiences often seem precisely the opposite of those which Gesell recommended. These patients seem to have felt, as children, that their own natural impulses and desires counted for little, or threatened others, and that they were forced to fulfill others' predetermined images and expectations (Laing, 1965; Bettelheim, 1967; White, 1963). We will discuss this matter more fully when we consider Bettelheim (Chapter 9), but here I would like to illustrate the point with a brief description of a nine-year-old boy whom I saw for a psychological evaluation. The boy found life very frightening and probably was on the verge of psychosis. The parents had not wanted a child, since they were in their 50s, and the mother had a number of physical ailments which made it taxing for her to care for him. Consequently, she wanted a good, well-disciplined boy—an adult, really—who would cause her no trouble. She tried to toilet-train him at six months of age, long before he showed any readiness to participate in the process. And when he began walking and vigorously exploring the world at one year, she became distressed; he was becoming a nuisance and "getting into things." She even perceived his behavior as abnormal. Because of her circumstances, then, she practically reversed Gesell's advice: She had a fixed image of the good child she wanted and was unable to accept and follow his natural inclinations. As a result, the boy developed an intense fear that any action he might take, unless approved by his parents beforehand, was extremely dangerous. He did not trust himself or his natural impulses.

There is some evidence, then, in support of Gesell's position—that it is desirable to respond to children's cues and inclinations as they follow an inner, biological schedule. However, there also is evidence which might argue partly against Gesell. In particular, research by Baumrind (1967) suggests that independent, self-reliant, and mature three- and four-year-olds have parents who demand a great deal of them. Baumrind thinks that these parents set tasks that are within their children's abilities, and to this extent the parents follow Gesell's recommendations. However, these parents also seem more demanding and controlling than Gesell might have thought necessary.

Perhaps philosophies such as Gesell's will never be completely proven or refuted by empirical evidence alone; too much may depend on one's own values. All the same, it would seem that we have much to gain by listening to Gesell. For while it is true that we must control, direct, and instruct our children to some extent, we usually seem to be in quite a hurry to do these things. What seems more difficult for us is to take the time to watch, enjoy, and appreciate our children as we give them a chance to do their own growing.

3

Ethological Theories:
Darwin,
Lorenz and Tinbergen,
And Bowlby

Ethology is the study of animal and human behavior within an evolutionary context. The person most identified with modern evolutionary theory is Darwin.

DARWIN AND THE THEORY OF EVOLUTION

Biographical Introduction

Charles Darwin (1809-1882) was born into a distinguished English family. His grandfather, Erasmus Darwin, was a renowned physician, poet, and philosopher, and his father also was a prominent physician. Young Darwin, in contrast, seemed headed for no great heights. As his father once said, "You care for nothing but shooting, dogs, and rat-catching, and you will be a disgrace to yourself and your family" (Darwin, 1887, p. 28).

Darwin studied medicine for a while, and then theology at Cambridge, but his performances were generally unexceptional. However, he had a winning personality and made a favorable impression on some of his professors at Cambridge, especially those who shared his interests in hunting and wildlife. One professor, John Henslow, recommended Darwin for the position of naturalist on the world-wide voyage of the H.M.S. Beagle, the voyage on which Darwin made observations which eventually led to his theory of evolution.

As Darwin examined fossils and variations among living species, he concluded that the various species had a common ancestor and that newer species had either died out or had changed to meet the requirements of their changing environments. If this conclusion were correct, then the common theological view on the origin of the species must be wrong; the species had not been created in a fixed and perfect form, but had evolved. Although the idea of evolution had been expressed before, its plausibility deeply distressed the religious Darwin. It seemed like "confessing murder" (Kardiner and Preble, 1961, p. 24).

Darwin was so troubled and uncertain about his theory that he did not publish it until 20 years after the voyage. In fact, he might never have published his theory at all had he not learned that Alfred Wallace was going to publish a similar theory. Since the theory was going to be made public anyway, Darwin wanted partial credit for it. Upon the recommendation of their colleagues, Darwin and Wallace presented their theory in 1858 under joint authorship. A year later, Darwin published his great work, *The Origin of Species*. Darwin continued developing his theory the rest of his life, and, despite the bitter reactions it produced, he became widely recognized for his monumental achievements. When Darwin died, he was buried in Westminster Abbey, next to Isaac Newton.

The Theory of Natural Selection

As mentioned, Darwin was not the first to propose a theory of evolution. In Darwin's day, biologists had been debating the views of Lamarck, who proposed that evolution occurred through the inheritance of acquired characteristics. For example, giraffes stretched their necks to reach the leaves on high trees and then passed along their lengthened necks to the next generation. Lamarck's theory, however, turned out to be wrong.

In the Darwin-Wallace theory, no new characteristics need be acquired during an individual's lifetime. In essence, Darwin's theory is as follows. Among the members of a species, there is endless variation, and among the various members, only a fraction of those who are born survive to reproduce. Thus, there is a "struggle for existence" during which the fittest members of a species live long enough to transmit their characteristics to the next generation. Over countless generations, then, Nature "selects" those who can best adapt to their surroundings—hence the term "natural selection" (Darwin, 1859, Chapters 3 and 4).

Darwin asked us to consider, for example, the situation of wolves (1859, p. 70). During seasons when prey is scarce, the swiftest and strongest wolves have the best chances for survival. They are therefore more likely than the others to live long enough to reproduce and pass on their traits—today we would say their genes—to the next generation. After many such seasons, the traits of speed and strength will become increasingly prevalent in the population of the species.

Evolution is an incredibly slow process; it usually produces noticeable changes only after numerous generations. Consequently, we do not usually get a chance to see evolution at work in any simple way. However, biologists did have such an opportunity in England in the mid-1800s. In Manchester, there lived numerous white moths who blended in well with the white trees in the area, making it difficult for birds to spot and kill them. Among the moths, there were a small number of dark ones (the products of mutations), who were easily detected by the predators. Consequently, only a few dark moths lived and reproduced. However, with industrialization, coal smoke darkened the trees, making the white moths easy prey. Now the dark ones had the best chance of surviving long enough to reproduce, and over the next 50 years their number increased from less than 1 percent to 99 percent (Ehrlich and Holm, 1963, pp. 126-30).

Social behavior and reason. Darwin's terms "struggle for existence" and "survival of the fittest" conjure up images of individuals in violent combat. In one sense, such images accurately reflect Darwin's views; for males of various species do compete for the females, as when deer battle during the spring. Such competitions ensure that the strongest males, rather than the weak and frail ones, will transmit their characteristics to the next generation (Darwin, 1871, p. 583).[1]

In another sense, however, terms such as the "struggle for existence" are misleading. For Darwin also wrote extensively on the importance of social instincts for group survival. Those animals that evolved danger signals and aided their fellow members must have had better chances for survival than those that did not. Similarly, those groups of early humans that banded together, cooperated, and looked out for the common good probably had a better chance of surviving (Darwin, 1871, p. 483, 500).

In humans, in addition, a crucial factor was the development of reason. Since humans are physically weaker and slower than many species, they must have relied on their intelligence and inventions (such as tools) to survive (1871, p. 444). Thus, the capacities for social behavior and reason, along with physical characteristics, must have undergone natural selection.

Evolution and embryology. Darwin (1859, p. 345) said that embryological findings fit well with the theory of evolution. In particular, he noted that the

---

[1] Darwin believed that the males often fight to the death (1871, p. 583). In most species, however, they probably do not, for they have evolved appeasement gestures which stop fights before they reach this point. Humans are a notable exception (Lorenz, 1963, pp. 129-38, 241).

embryos of most species are highly similar in their early forms, perhaps revealing their descent from a common ancestor. The strongest argument of this kind was made by Haeckel, who in the late 1860s proposed that *ontogeny recapitulates phylogeny.* That is, the development of an individual organism (ontogeny) repeats in an abbreviated way the evolutionary history of its species (phylogeny). The recapitulation argument has always seemed somewhat fanciful, but biologists still consider it a fair working model. For example, before mammalian embryos take on their distinctive shapes, they resemble the young embryos of fish, amphibians, reptiles, and birds—in that order, the same order in which the species appeared on the earth, as judged by fossil evidence (Waddington, 1962, p. 99).

Evaluation

Today Darwin is considered correct but incomplete. He was right in pointing out that there is enormous variation within species and that species change because only some members survive long enough to reproduce. However, it was only after the work of Mendel and others that we began to understand how genetic combinations and mutations produce variations and how traits are transmitted. Thus, the science of genetics helped round out Darwin's theory.

Darwin believed, as we have seen, that natural selection applies not only to physical characteristics (such as coloring) but also to various kinds of behavior. Thus, Darwin was the first ethologist—the name given to biologists who study animal behavior from an evolutionary perspective. We will now review some of the ideas of modern ethologists and then look at the applications of these ideas to the study of human development.

MODERN ETHOLOGY: LORENZ AND TINBERGEN

Biographical Introduction

Konrad Lorenz is often called the father of modern ethology. He has not necessarily made more discoveries than other ethologists, but his bold, vivid, and often humorous writing style has done much to call attention to this new field.

Lorenz was born in 1903 in Austria, where he still lives. His father was a prominent physician who wanted Lorenz to become a doctor too, so Lorenz dutifully earned a medical degree. However, he never lost his boyhood enthusiasm for the study of nature and wildlife, and he next studied zoology at the University of Vienna, earning a Ph.D. in this field. Lorenz began his studies in ethology in the early 1930s, when he became convinced that one can see the landmarks of evolution in the innate behavior patterns of animals just as surely as in their physical characteristics (Tanner and Inhelder, 1971, p. 28). He has made many of his observations on his own large estate, where many species of wild animals freely roam.

Niko Tinbergen has worked quietly in Lorenz's shadow. Despite this, ethologists consider his work equally substantial. Tinbergen was born in 1907 at the Hague, the Netherlands, and like Lorenz was fascinated by animals and wildlife as a boy. In school, Tinbergen's work was erratic; he did well only in subjects that interested him, and he struck many of his teachers as a lazy youngster whose primary enthusiasm was for sports. Nevertheless, Tinbergen went on to earn a Ph.D. degree in biology at the University of Leiden in 1932 and began doing brilliant ethological studies. His research was interrupted during World War II when the Germans put him in a prison camp for protesting the dismissal of Jewish professors at the university. During his imprisonment, Tinbergen wrote on ethology as well as stories for children. Since 1947 he has been a professor at Oxford. In 1973 Tinbergen and Lorenz, along with a third eminent ethologist, Karl Von Frisch, won the Nobel Prize in Biology (Baerends *et al.*, 1975).

Methodological Approach

Ethologists are convinced that we can only understand an animal's behavior if we study it in its natural setting. Only in this way can we watch an animal's behavior patterns unfold and see how they serve in the adaptation of the species. We cannot, for example, understand why birds build nests where they do unless we see how such behavior protects the species from predators in the natural environment. Psychologists who only study animals in their laboratories miss out on a great deal. In such captive situations, many species do not even reproduce, and one frequently has no opportunity to observe their nesting, mating, territorial, or parental behavior.

When an ethologist studies a new species, his or her first step is simply to get to know the species as well as possible. That is, the ethologist engages in *naturalistic observation;* the ethologist observes an animal's characteristic behavior and then compares the behavior with that of other species. Only after ethologists have gathered a great deal of descriptive material do they attempt experiments to test their ideas or try to formulate general laws.

Instinctive Behavior

Ethologists are interested in instincts. In everyday language, we casually refer to any unlearned behavior as "instinctive" or as an "instinct," but ethologists consider instincts as a special class of unlearned behavior.

An instinct, in the first place, is *released by a specific external stimulus.*

This is the case, for example, in the rescuing behavior of the chicken. The hen appears to respond any time her chicks are in danger, but a closer examination reveals that she is actually reacting to a very specific stimulus—the chick's distress call. This point has been demonstrated in an experiment by Brückner (cited in Tinbergen, 1951). When chicks were tied to a post and hidden behind a screen, the mother still rescued them because she heard the distress calls. When, however, the screen was removed and the struggling chicks were covered by a glass enclosure—so the hen could see her chicks in trouble but could not hear them call—she ignored them. She needs to hear the specific distress call.

Similarly, a specific stimulus releases the tendency to fight in the male three-pronged stickleback fish (Tinbergen, 1951, p. 28). In the spring the adult male establishes a territory, builds a nest, and courts females. He also develops a bright red spot on his belly. When other sticklebacks enter his territory, he may or may not fight—depending on the sight of a specific stimulus, red on the invader's belly.

If a pregnant female enters his territory, his behavior is different. When he moves toward her, she tilts her body upward, exposing her full belly, and this stimulus causes him to go into a zig-zag dance. This dance is a signal for her to approach, and when she does, he returns to the nest. Once there, his behavior signals for her to enter. This intricate ritual continues until the eggs are fertilized, with each component of the ritual governed by a specific releasing stimulus (Tinbergen, 1951, pp. 48-50).

Specific releasing stimuli also determine the reactions of the young to their parents. For example, a young pheasant will rush for cover only when it hears its parent's warning call. Similarly, a young jackdaw bird will follow its parent into the air only when the parent takes off at a certain angle and speed (Lorenz, 1935, pp. 157-61).

Instincts also are *species-specific,* which means that the particular behavior patterns are found only in members of a specific species. The behaviors always include some *fixed action pattern,* some stereotyped motor component. Fighting gestures, courtship behavior, and modes of following always contain some fixed aspect.

Not every part of an instinct, however, must be of a fixed nature. For example, peregrine falcons engage in free flight when searching for prey. There is no rigid pattern to their search; they glide around in various areas where past experience has taught them to look. But once they spot their prey (e.g., a starling bird), their actions do become stereotyped. Their swoop and their manner of catching their prey constitute a fixed action pattern (Lorenz, 1952a, p. 306).

At the same time, some parts of a motor performance may be less fixed than it first appears. Bowlby (1969) points out that the falcon must be able to adjust its flight if it is to intercept moving prey. He thinks the falcon possesses, in addition to stereotyped movements, a "goal-corrected" feedback mechanism by which it alters its movements in accordance with those of the prey, much as a missle is programmed to do to intercept a moving target. Lorenz has not commented on such a mechanism, but his description of young jackdaws' following behavior suggests that after a time they develop a similar capacity to maintain contact with a parent in flight (Lorenz, 1935, p. 159).

The fixed action pattern also has an underlying *drive component,* an inner urge to engage in the instinctive behavior. Consequently, if the behavior has not been released for a long time, the drive behind it can build up to the point that less specific stimuli will suffice, as when males court females who lack the specific releasing stimuli. In some cases, the internal pressure for release builds up to such a high pitch that the fixed action pattern goes off "in a vacuum" (Lorenz, 1963, pp. 52-53).

Finally, instincts, as the products of evolution, have some *survival value for the species.* However, as Freedman (1971) observes, it is often all too easy to create plausible-sounding explanations of a behavior's adaptive value. What is really needed is research on the question. For example, Tinbergen (1965) wondered why herring gulls remove their egg shells after hatching. After all, this requires them to spend time away from the nest, endangering their young. He hypothesized that the glistening of the shells in the sun attracts predators, and he conducted an experiment to see if this were so. He scattered some egg shells in one area and found that, in fact, predators investigated this area much more than a comparable area lacking the shells.

As mentioned earlier, instincts are distinguished from other forms of unlearned behavior. An instinct seems different from a general drive such as hunger, because the hunger drive is found in many species (it is not species-specific). Instincts also seem different from reflexes. Instincts may contain reflexes, but they also may be more complex. For example, the stickleback's zig-zag dance must involve many reflexes. Also, a reflex, such as an eye-blink, can be released by many stimuli—wind, noise, dust, bright lights, and so on. There does not seem to be a specific external releaser.

Imprinting

In many instances, an animal's responsiveness to specific releasers is innate or inborn. In many other instances, however, crucial aspects of the stimuli which release social instincts are acquired during a susceptible period early in life. For example, many species of young birds and mammals direct their following responses toward their mothers because she was the first object they saw and followed during a specific time in infancy. When Greylag goslings were raised by Lorenz instead of by their mothers, they took him for their "mother." They energetically followed him about in single file wherever he went, ignoring adult geese. According to Lorenz (1935, p. 124), the goslings *imprinted* on him; that is, they formed, during an early period, an "imprint" of him as the object of their attachment.

Although Lorenz was not the first to observe imprinting, he was the first to state that it occurs during a *critical period*. This means that the young animal will form an attachment to an object only if it is exposed to and follows that object during a specific time early in life. If it is exposed to an object prior to, or after the critical period, no attachment is formed. And once the critical period has passed, it may be impossible to induce the animal to attach itself to another kind of object (Lorenz, 1935, p. 127).

Lorenz found that species differ with respect to the range of objects on which they will imprint. Greylag goslings seem to imprint on almost anything that moves (some even have imprinted on boats). Mallard ducklings, in contrast, are more finicky. Lorenz found that they would only imprint on him if he stooped below a certain height and made a quacking sound as he moved about. Mallards, then, have an innate schema of certain aspects of the proper parent— the parent must move, be of a certain height, and make a certain sound. Imprinting only fills in the rest of the visual pattern (Lorenz, 1935, p. 135; 1952b, pp. 42-43).

Imprinting can determine not only the following response in the young but later social behavior as well. In particular, early imprinting can affect later sexual preferences, as Lorenz also learned from personal experience. One of his neighbors hand-raised an orphan jackdaw bird which imprinted on humans, and when this bird reached sexual maturity it courted Lorenz. It attempted to seduce him in typical jackdaw fashion, by inserting crushed worms into Lorenz's mouth. When Lorenz shut his mouth, the bird put the worms into his ear. This bird, then, having been exposed only to humans during its early life, focused on humans as the objects of its later sexual advances. The critical period for sexual imprinting may differ from that for parental imprinting, but it too occurs very early, long before sexual behavior actually emerges (Lorenz, 1935, p. 132; 1937, p. 280; 1952b, pp. 135-36).

So far, we have been discussing the formation of social attachments, parental and sexual. Imprinting-like processes also can govern other kinds of learning, including the learning of territorial maps, food preferences, and songs. For example, the song of the White-crowned sparrow always has a certain basic structure, but some details vary from one geographical location to the next. These details are determined by the song the bird hears during an early critical period (Marler and Tamura, 1964).

Nevertheless, most research on imprinting has focused on the formation of social attachments, especially on the early following response. Some of this research has raised questions about some of Lorenz's initial formulations. For one thing, Lorenz (1935) initially proposed that parental imprinting is always supra-individual. That is, he believed that the young imprint on a particular species of parent but not on an individual parent. However, other ethologists have observed that some species do imprint on individual parents (Bateson, 1966), and Lorenz has since modified his position. He now thinks that goslings, for example, first fixate on a species and then on an individual parent—all within the first two days of life (Lorenz, 1965, p. 57).[2]

Other research, especially that conducted in laboratories, has cast some doubt on Lorenz' hypothesis that imprinting is completely irreversible. There is some evidence that some birds can learn to follow objects different from those that they were imprinted upon. Still, this learning is much slower than the original imprinting, and the birds still seem to prefer their imprinted object (Bateson, 1966).

Some of the best laboratory research on imprinting has been carried out by Eckhard Hess (1962; 1973, Ch. 4) at the University of Chicago. Hess, who has primarily studied Mallard ducklings, has been especially interested in the events associated with the boundaries of the critical period. He has found that imprinting *begins* when the ducklings are able to exert some effort in making a following response, and that the strength of imprinting is a function of the effort exerted. If, during the critical period, ducklings must run up inclines or climb over hurdles to follow a decoy, their later following response to the decoy will be all the stronger. This finding, however, is more limited than Hess originally thought, for some birds can imprint on a parent without following it at all (Hess, 1973, p. 113). Hess has suggested that the critical period *ends* with the onset of the fear response; the young then avoid any new or strange object and search instead for the imprinted parent. This hypothesis has gained fairly wide acceptance (Bateson, 1966).

Ethologists also have wondered about the adaptive value of imprinting. It appears to have evolved as a particularly early, strong attachment mechanism in those birds and mammals (including deer, sheep, and buffalos) that live in groups, move about soon after birth, and are under strong predator pressure. In these species, the quick formation of a following response insures that the young will follow an escaping parent in times of danger (Freedman, 1974, p. 19).

Still, one can ask why imprinting is necessary at all. Why, that is, have not the young of all vulnerable species evolved built-in parental schemas? One suggestion is that adult appearances often vary and change over time, so the young can benefit from some opportunity to discover for themselves who their parents are (Bateson, 1966).

Although imprinting is an especially early, rapid process in some species, some form of imprinting may occur in a wide range of species, including primates such as chimpanzees. Young chimpanzees do not show much concern over whom they are with until they are about three or four months old. Then they develop a marked preference for their mother (or foster parent) and become

[2] Sexual imprinting, in contrast, does usually seem to determine only the species that the young will later court, not the particular member of the species. One can only be sure that the young will not court its imprinted parent; this is somehow excluded (Lorenz, 1935, p. 132).

distinctly wary of other adults. After this, they stay fairly near her, returning to her from time to time, and if she should signal that she is about to depart, they rush over and climb aboard. Thus, chimps clearly attach themselves to a particular object during a certain period in life (Bowlby, 1969, p. 190, 196). A similar process may occur in human children, as we shall discuss presently.

Evaluation

Ethology has given us a number of new, interesting concepts and new ways of looking at behavior. At the same time, it has drawn heavy fire from many psychologists, particularly from American psychologists who have criticized it for ignoring the role of learning and experience (e.g., Schnierla, 1960; Riess, 1954). Such attacks, however, seem largely misdirected. For ethologists do not, as it is often supposed, claim that behavior is entirely innate and that learning and experience count for nothing. Imprinting, after all, is a kind of learning. It is a process by which the stimulus properties of social releasers become filled in through experience. Imprinting is not, to be sure, the kind of learning that Locke, Pavlov, or Skinner talk about. It is a remarkably permanent learning that occurs only during a maturationally determined critical period. Still, it is a kind of learning.

Some psychologists have tried to refute the concept of instinct by showing that instincts do not develop properly if animals are deprived of key experiences. For example, Riess (1954) showed that rats deprived of experiences with sticks failed to show the nest-building instinct. Such experiments, however, miss the ethologists' point. Ethologists recognize that instincts have evolved because they have been adaptive within certain environments and that instincts need the right environment to develop properly. The environment is important. All that ethologists claim is that instinctive behaviors have a large innate component and that, given the environment to which the instinct is preadapted, the instinct will emerge without elaborate conditioning or learning.

Ethology is still a very young science, so we can still debate its ultimate fruitfulness. For our purposes, the basic question is: How useful is it for understanding human development? We will now turn to a major effort to apply ethological concepts to human development—that of John Bowlby.

AN ETHOLOGICAL VIEW OF CHILD DEVELOPMENT: BOWLBY'S THEORY

Biographical Introduction

John Bowlby was born in 1907 in London, where he still lives and works. He has taught in a progressive school for children, received medical and psychoanalytic training, and, since 1936, has been primarily involved in child guidance work. In 1936 Bowlby became concerned about the disturbances of children raised in institutions. Children who grow up in nurseries and orphanages, he found, frequently show a variety of emotional problems, including an inability to form intimate and lasting relationships with others. It seemed to Bowlby that such children are unable to love because they missed the opportunity to form a solid attachment to a mother-figure early in life. Bowlby also observed similar symptoms in children who grew up in normal homes for a while but then suffered prolonged separations. These children seemed so shaken that they permanently turned away from close human ties. Such observations convinced Bowlby that one cannot understand development without paying close attention to the mother/infant bond. How is this bond formed? Why is it so important that, if disrupted, severe consequences follow? In his search for answers, Bowlby turned to ethology (Tanner and Inhelder, 1971, pp. 25-27).

Theory of Attachment: Overview

Bowlby says that we can only understand human behavior by considering its *environment of adaptedness,* the basic environment in which it evolved (1969, p. 58). Throughout most of human history, humans probably moved about in small groups, searching for food, and often risking attack by large predators. When threatened, humans, like other primate groups, probably cooperated to drive off the predators and to protect the sick and young. To gain this protection, human children needed mechanisms to keep them close to their parents. That is, they must have evolved *attachment behaviors*—gestures and signals which promote and maintain proximity to caretakers (Bowlby, 1969, p. 182).

One obvious signal is the baby's cry. The cry is a distress call; when the infant is in pain or is frightened, he or she cries and the parent is impelled to rush over to see what is wrong. Another attachment behavior is the baby's smile; when a baby smiles into a parent's eyes, the parent feels love for the baby and enjoys being close. Other attachment behaviors include babbling, grasping, sucking, and following.

Bowlby suggests that the child's attachment develops along the following lines. Initially, babies' social responsiveness is indiscriminate. For example, they will smile at any face or cry for any person who leaves. However, between about three and six months of age, babies narrow their responsiveness to a few familiar people and develop a clear-cut preference for one person in particular. They then become wary of strangers. This formation of an exclusive attachment parallels the imprinting process in other species.

After this, babies become more mobile and crawl about and take a more active role in keeping the attachment-figure nearby. They monitor this parent's whereabouts, and any sign that the parent might suddenly depart releases following on their part.

Throughout his discussions, Bowlby speaks of attachment behaviors (such as crying, smiling, and following) as "instinctive." However, he uses the term instinct in a purposely loose sense. He means that such behaviors are basically innate, have a fairly typical pattern in almost all members of the species, and have an adaptive value for the species (1969, p. 136). Bowlby does not try to show how each behavior pattern is released by a specific external stimulus, although he does talk about releasers when it is convenient to do so.

Let us now look more closely at the phases of attachment.

Phases of Attachment

Phase 1 (birth to three months): Indiscriminate Responsiveness to Humans. During the first months of life, babies show various kinds of responsiveness to people, but this responsiveness is unselective; babies react to most people in fairly similar ways.

Right after birth, babies like to listen to human voices and to look at human faces (Fantz, 1961; Freedman, 1974, p. 23). A recent study, for example, indicates that babies who are only 10 minutes old prefer the face to other visual stimuli; they move their heads farther to follow an accurate model of a face than to follow scrambled faces or a blank sheet (Jirari, in Freedman, 1974, p. 30). To ethologists such as Bowlby, this preference suggests a genetic bias toward a visual pattern which will soon release one of the most powerful attachment behaviors, the *social smile.*

During the first three weeks or so, babies will sometimes emit eyes-closed smiles, usually as they are about to fall off to sleep. These smiles are not yet social; they are not directed at people. At about three weeks, babies begin smiling at the sound of a human voice. These are social smiles, but they are still fleeting (Freedman, 1974, pp. 178-79).

Beginning at about five weeks, the most intense social smiling begins (Freedman, 1974, pp. 180-81). Babies smile happily and fully at the sight of a human face, and their smiles include eye contact. One can tell when such visual smiles are about to start. About a week beforehand, the baby starts to gaze intently at faces, as if studying them. Then the baby breaks into open smiles. This is often an electrifying moment in a parent's life; the parent now has "proof" of the baby's love. The sight of one's baby looking deeply into one's eyes and smiling causes a feeling of love to well up from within. (Even if you are not a parent, you might have had a similar feeling when any baby smiled at you. You cannot help but smile back, and you think that you and the baby share a special bond.)

Actually, until three months or so, the baby's smiles show no personal preferences. Babies will smile at any face, even a cardboard model of one. The main stipulation is that the face be presented in the full or frontal position. A profile is far less effective. Also a voice or a caress is a relatively weak elicitor of smiling during this stage. It seems, then, that the baby's social smile is released by a fairly specific visual stimulus (Bowlby, 1969, pp. 282-85; Freedman, 1974, p. 187).

In Bowlby's view, smiling promotes attachment because it maintains the proximity of the caretaker. When the baby smiles, the caretaker enjoys being with the baby; the caretaker "smiles back, 'talks' to him, strokes and pats him, and perhaps picks him up" (Bowlby, 1969, p. 246). The smile itself is a releaser which promotes loving and caring interaction—behavior which increases the baby's chances for health and survival.

At about the time that babies begin smiling at faces, they also begin *babbling* (and cooing and gurgling). They primarily babble at the sound of a human voice, and, especially, at the sight of a human face. As with smiling, babbling is initially unselective; babies will babble when most any person is around. The baby's babbling delights the caretaker, prompting the caretaker to talk back. "Babbling, like smiling, is a social releaser [which] has the function of maintaining a mother-figure in proximity to an infant by promoting social interaction between them" (Bowlby, 1969, p. 289).

Crying also results in proximity between caretaker and child. It is like a distress call; it signals that the baby needs help. Babies cry when they are in pain, hungry, cold, or experience discomfort. They even cry when a person at whom they had been looking leaves their field of vision, although during the first weeks the particular person in question matters little. Babies also will permit most any person to quiet them through rocking or by attending to their needs (1969, pp. 289-96).

Proximity also is maintained by the baby's *holding on.* The newborn is equipped with two holding responses. One is the grasp reflex; when any object touches the baby's open palm, the hand automatically closes around it. The other is the Moro reflex, which occurs when babies either are startled by a loud noise or when they suddenly lose support (as when one holds their head from underneath and then suddenly lets it drop). Their response is to spread their arms and then to bring them back around the chest. The action looks as if the baby were embracing something. In earlier times, Bowlby thinks, these reflexes served the purpose of keeping hold of the parent who carried them about. If, for example, a mother saw a predator and suddenly ran, the chances were that the baby had a grasp of some part of her with the hand. And if the baby lost hold, he or she would embrace the mother again (1969, p. 278).

Babies also are equipped with *rooting* and *sucking* reflexes. When their cheek is touched, they automatically turn their head to the side from which the stimulation came and then "root" or grope until their mouth touches something, which they then suck. The rooting and sucking reflexes obviously aid breast-feeding, but Bowlby also regards them as attachment patterns because they bring the baby into interaction with the mother (1969, p. 275).

Phase 2 (three to six months): Focusing on Familiar People. Beginning at about three months, the baby's behavior changes. For one thing, many reflexes—

including the Moro, grasp, and rooting reflex—drop out. But more importantly for Bowlby, the baby's social responses begin to become much more selective. Between three and six months, babies gradually restrict their smiles to familiar people; when they see a stranger, they simply stare (Bowlby, 1969, p. 287, 325). Babies also become more selective in their babbling; by the age of four or five months, they coo, gurgle, and babble only in the presence of people they recognize (1969, p. 289). Also, by this age (and perhaps long before), their crying is most readily quieted by a preferred figure (1969, p. 279, 300). By five months, finally, babies begin to reach for and grasp parts of our anatomy, particularly our hair, but they do so only if they know us (1969, p. 279).

During this phase, then, babies narrow their responsiveness to familiar people. They usually prefer two or three people—and one in particular. They most readily smile or babble, for example, when this person is near. This principal attachment-figure is usually the mother, but it doesn't have to be. It can be the father or some other caretaker. Babies seem to develop the strongest attachment to the one person who has most alertly responded to their signals and who has engaged in the most pleasurable interactions with them (1969, pp. 306-16).

Phase 3 (six months to three years): Active Proximity-seeking. At about six months of age, babies show a deep concern for the attachment-figure's presence and will often cry when he or she starts to leave. Earlier, they might have protested the departure of anyone at whom they had been looking; now, however, it is primarily this one person's absence that upsets them (1969, p. 295).

Babies, however, are not doomed to crying and helpless waiting. By seven months or so they usually can *crawl* and therefore can *actively follow* a departing parent. At this point, their behavior begins to consolidate into a *goal-corrected* system. That is, babies monitor the parent's whereabouts, and if the parent starts to leave, they urgently follow, "correcting" or adjusting their movements until they regain proximity. When they get near the parent once again, they typically reach up with their arms, gesturing to be picked up. When held, they calm down again (1969, p. 252).

Babies also begin to *call out* in a new way, and calling also takes on a goal-corrected quality. That is, babies alter their calls according to their estimates of the parent's location and movements. If the parent is far or departing, they call loudly; if the parent is near or approaching, they call less urgently (1969, p. 252).

In these behavior systems, the attachment-figure's departure is considered a stimulus which activates proximity-seeking. Babies will make the most concerted efforts to regain contact when a parent departs suddenly, rather than withdrawing slowly and quietly, or when they are in unfamiliar surroundings. Other eliciting stimuli include events that frighten the child, such as a loud noise or sudden darkness (1969, pp. 256-59).

Babies, of course, often move away from attachment-figures as well as toward them. This is particularly evident when they use the caretaker as *a base from which to explore.* If a mother and her one- or two-year-old child enter a park or playground, the child typically remains close for a moment or two, and then ventures off to explore. However, the child periodically looks back, exchanges glances or smiles with her, and even returns to her from time to time before venturing forth again. Thus, the child initiates brief contacts to assure himself or herself that proximity is maintained. If the mother fails to respond, or threatens to leave, the child resorts to more urgent or clinging tactics (1969, p. 208).[3]

Bowlby emphasizes that the child's ability to maintain contact, while developing, remains incomplete until at least three years of age. Not until this age can children adjust their movements to keep up with a parent when accompanying him or her on a walk. As with the young in some other species, a complete goal-corrected following system takes a while to develop (1969, p. 254).

During this phase, attachment becomes increasingly intense and exclusive. The intensity of the baby's attachment is seen, as mentioned, when the baby cries over the parent's departure. This is often called *separation anxiety.* Observers have also noted the intensity with which the baby *greets* the mother after she has been away for a brief time. When she returns, the baby typically smiles, reaches to be picked up, and, when in the mother's arms, hugs her and crows with delight. The mother, too, displays her happiness at the reunion (1969, p. 200).

The new exclusiveness of the baby's attachment is most evident at about seven or eight months of age, when the baby shows a *fear of strangers.* This reaction ranges from a slight vigilance to outright cries at the sight of a strange person. The stronger reactions usually occur when the baby feels ill or is in an unfamiliar setting. When alarmed, the baby *clings* to the parent, and if the parent should then try to hand the baby over to a stranger, the baby will cling all the more tightly (1969, p. 321, 279).

Phase 4 (three years to the end of childhood): Partnership Behavior. Prior to the age of two or three years, children are only concerned with their own need to maintain a certain proximity to the caretaker; they do not yet consider the caretaker's plans or goals. For the two-year-old, the knowledge that mother or father is "going next door for a moment to borrow some milk" is meaningless; the child simply wants to go too. The three-year-old, in contrast, has some

[3] In one study, it was found that two-year-olds rarely ventured more than 200 feet from the mother. The mother, too, seems to intuitively set a distance beyond which she will not permit her child to stray. If children do so, mothers retrieve them (Bowlby, 1969, p. 258, 240).

understanding of such plans and can visualize the parent's behavior while he or she is away. Consequently, the child is more willing to let the parent go. The child begins acting more like a partner in the relationship.

Bowlby admits (1969, p. 387) that little is known about phase 4 attachment behavior, and he has little to say about attachments during the rest of life. Nevertheless, he feels that they remain very important. Adolescents break away from parental dominance, but they form attachments to parental substitutes; adults consider themselves independent, but they seek proximity to loved ones in times of crisis; and older people find that they must increasingly depend on the younger generation (1969, p. 207). In general, Bowlby says, being alone is one of the great fears in human life. We might consider such a fear silly, neurotic, or immature, but there are good biological reasons behind it. Throughout human history, humans have best been able to withstand crises and face dangers with the help of companions. Thus, the need for close attachments is built into our nature (Bowlby, 1973, p. 84, 143, 165).

Attachment as Imprinting

Now that we have examined the child's attachment in some detail, we are in a position to appreciate Bowlby's thesis that it follows a course similar to imprinting in animals. Imprinting, you will recall, is a process by which young animals learn the objects of their social responses. They begin with a willingness to follow a wide range of objects, but this range quickly narrows. At the end of the imprinting period, young animals are often attached to a single object— usually the mother. Thereafter, the fear response limits the ability to form new attachments.

In humans, we can observe a similar process, although it develops much more slowly. During the first weeks of life, babies cannot actively follow objects through locomotion, but they do direct social responses toward people. They smile, babble, hold on, cry, and so on. At first, they direct their responses toward anyone. However, by six months of age they have focused their attachment to a few familiar people and to one person above all. They then show a fear of strangers. As with animals, then, the fear of strangers limits the possibilities for forming new attachments. Thereafter, babies are keenly aware of their principal attachment-figure's whereabouts and develop a goal-corrected following system for maintaining proximity to this person.

Effects of Institutional Care

Institutional deprivation. As we mentioned in the introductory remarks, Bowlby turned to ethology as a way of accounting for the damaging and seemingly irreversible effects of institutional deprivation. He was particularly struck by the inability of many institutionally reared children to form deep attachments later in life. He called these individuals "affectionless characters"; such individuals use people solely for their own ends and seem incapable of forming a loving, lasting tie to another person (Bowlby, 1953). Perhaps these people, as children, lacked the opportunity to "imprint" on any human figure—to form a loving relationship with another. Having missed the opportunity to develop the capacity for intimate ties during the normal early period, their relationships remain shallow in adult life.

The conditions in many institutions do seem unfavorable for the formation of intimate human ties. In many institutions, babies receive care from several nurses who can meet their physical needs but who have little time to interact with them. Frequently, no one is around to heed the babies' cries, to return their smiles, to talk to them when they babble, or to pick them up when they desire. Consequently, it is difficult for the baby to establish a strong bond to any particular person.

If a "failure to imprint" explains the effects of institutional deprivation, there should be a critical period beyond which these effects cannot be reversed. Researchers, however, have had trouble pinpointing the precise ages of any such critical period. Bowlby's discussion of imprinting (1969, pp. 222-23; see also 1953, p. 58) suggests that the critical period ends with the onset of the fear response, just as it does with other species. This would place the end of the critical period at eight or nine months, the age by which almost all babies have shown some fear of separation from the mother as well as a fear of strangers. In fact, there does seem to be some evidence for this age as the termination point for the appropriate development of some social behavior. In particular, babies who lack human interaction up to this point seem to be permanently retarded in vocalization (Ainsworth, 1962). However, it also appears that therapeutic interventions can make up for most social deficits up to the age of 18 to 24 months. In one view, institutional deprivation puts babies in "cold storage," slowing up social growth and extending the critical or sensitive period for attachments into the second year. After this point, babies lacking in human interaction may never develop normally (Ainsworth, 1973).

Separations. Although Bowlby has been interested in "failures to imprint," he has been even more interested in cases where the child forms an attachment and then suffers a separation. For example, the child may need to be placed in a hospital for a few weeks because of a physical illness.

According to Bowlby and Robertson (Bowlby, 1969, Ch. 2), the effects of separation typically run the following course. First, babies *protest*; they cry and scream and reject all forms of substitute care. Second, they go through a period of *despair*; they become quiet, withdrawn, and inactive, and appear to be in a deep state of mourning. Finally, a stage of *detachment* sets in. During this period, the child is more lively and may accept the care of nurses and others. The hospital staff may think that the child has recovered. However, all is not well. When the mother returns, the child seems not to know her; the child turns away and seems to have lost all interest in her.

Fortunately, most children do reestablish their tie to the mother after a while. But this is not always the case. If the separation has been prolonged, and if the child loses other caretakers (e.g., nurses), the child may give up on people altogether. The result here, too, is an "affectionless character," a person who no longer cares for others in any deep way.

The effects of separation also vary with the child's age. Separations seem most disruptive between six months and a year, right after the baby has formed an attachment. In contrast, separations may be less damaging after the age of three or four years, during Bowlby's fourth phase, when the child has become better able to tolerate the mother's absence and can understand the reasons for it (Ainsworth, 1973).

Implications for Child-Rearing

Bowlby believes, then, that babies have innately determined responses and signals which keep the parents nearby and foster an increasingly intimate relationship with them. These responses include smiling, babbling, crying, holding, sucking, and following. When, for example, the baby smiles or babbles, the mother interacts with the baby, and baby and mother begin to enjoy each other. When the baby cries, the mother is gratified by her ability to provide comfort. Through such interactions, babies increasingly come to single out and to love this person who provides so much care and pleasure.

If the baby's signals and responses are part of a biological ground plan for social development, development should proceed in the best way when the parent is responsive to them. The research of one of Bowlby's followers, Mary Ainsworth, suggests that this is so. We already have mentioned one of Ainsworth's studies (Bell and Ainsworth, 1972) in our discussion of Gesell. This study suggested that when mothers consistently and promptly respond to their babies' cries, the babies actually cry relatively little and are quite independent at one year of age. Apparently, when babies develop the sense that they can elicit the mother's care when necessary, they have less reason to worry about her presence.

Other research has produced similar results. Ainsworth (1967, 1973) has found that some mothers rather consistently ignore their babies' signals—their cries and their signs that they want to play. These babies tend to become isolated. Other mothers respond warmly, but when they want to, not according to their babies' own timing. These babies tend to become overly attached. They seem uncertain that the mother will respond when needed and become overly upset and helpless when left alone. Some mothers, finally, respond both warmly *and* in accordance with the babies' own signals. These babies, at one year, are "securely attached." They like to have the mother nearby, but they also can leave the mother's lap and independently explore a new situation.

The ethological position, then, is that parents should follow their inner tendencies to respond to their babies' cues (Ainsworth, 1973; Bowlby, 1969, p. 357). This position is identical to Gesell's. The similarity is based on the shared conviction that evolution has provided the child with innate responses that can be trusted to lead to healthy development.

Why is it that parents do not automatically follow their children's lead? The answer varies from parent to parent, but any general answer must include a consideration of our advanced status in the evolutionary cycle. The human adult, in comparison to other animals, is very much a product of education and cultural training. As a result, it is easy to assume that our own educational practices count for everything and that innate tendencies are insignificant. We think we can raise the child in any way we want—according to our own ideas— and we tend to overlook the extent to which children are biologically prepared to guide us with respect to the experiences they need.

Effects of institutionalization. Bowlby, as we have seen, was among the first to alert us to the potentially harmful effects of institutional deprivation and separations. Fortunately, people seem to be becoming aware of Bowlby's findings. For example, some hospitals are aware of the effects of separation and permit mothers to room-in with their babies.

Bowlby's work also has implications for adoption and foster placements. If we must move a child from one home to another, we should consider the baby's stage of attachment. When possible, it would seem wisest to place a baby in a permanent home during the first few months of life, before the baby begins to focus his or her love onto a single person. Between six months and three or four years, separations are likely to be the most painful. During this time, babies still lack both the independence to adjust to separations and the cognitive capacities to understand the reasons behind them (Ainsworth, 1973).

We might take comfort in the knowledge that most children do not have to grow up in impersonal institutions or suffer separations or foster placements. Such complacency, however, may be unfounded; the effects of institutionalization may be more widespread than we have assumed. In a recent issue of the *Journal of Pediatrics* (1977), Lozoff and her associates suggest that one kind of institutionalization is the rule—not the exception—in child development. This is the *institutionalized care of the newborn.*

Lozoff's paper owes much to Bowlby. She notes that throughout most of human history, babies began life in an environment of close physical contact with their mothers. In this "environment of adaptedness," babies evolved characteristics and responses that facilitate attachment to a caretaker right from the start. Newborns, Lozoff points out, are wide-eyed and alert an hour after birth, prefer looking at a human face and hearing a human voice, stop crying when lifted to an adult's shoulder, and strike most parents as beautiful. Such responses and characteristics immediately stir feelings of love and responsiveness in the mother. She loves this baby who looks at her intently, who takes comfort from her hold, and who looks so beautiful. She feels these things, that is, when she is in close contact with her baby, as she has been throughout most human history.

In the modern hospital, however, this is not the case. Babies are separated from their mothers for long periods of time; they are kept in nurseries and fed on a routine four-hour-schedule. Nurseries are designed to prevent the spread of infections, but they also may be placing a great strain on the beginnings of the mother-child relationship.

In support of her position, Lozoff marshals a substantial amount of evidence that indicates that development proceeds along better lines when mothers and babies are permitted just a few extra hours of daily contact during their hospital stay. When extra care is permitted, babies cry less in the hospital and mothers feel more confident in their handling of them. During the next few months, the mothers continue to feel more confident and to soothe, kiss, and smile at their babies more frequently. They also are more likely to breast-feed their babies and to talk to them in gentler tones; child abuse is rare. For their part, babies smile and laugh more, and cry less.

These findings suggest, then, that we should allow much more mother-infant contact in hospitals than we ordinarily do. In this day of widespread complaints about disturbances in parenting and people's inability to maintain lasting relationships, it would seem wise to do whatever we can to see that attachments get off to the best possible start.

Evaluation

Bowlby and the ethologists have given us a new way of looking at human development. After reading Bowlby, we can no longer pass quickly over behaviors such as a baby's cries or smiles without wondering about their evolutionary significance. We think, for example, about the baby's cries as similar to distress calls in other species. Similarly, we see the baby's smile as a behavior pattern that generates love and care in the parent, thus helping the species survive.

Even "childish" behaviors begin to look different from an ethological perspective. If, for example, a two-year-old girl is brought into an unfamiliar place by her mother, the girl might repeatedly turn to make sure her mother is present, and if she should lose sight of her mother, she might cry out. To many adults present, this child might seem immature and insecure; surely the mother ought to shape her up as soon as possible. Ethological considerations, however, suggest that this girl is following her instincts. As with other species, she has innate signals, such as cries and a following response, which help maintain proximity to the mother in unfamiliar settings. Such signals have become part of her nature because they have aided in the survival of the species. If, over the centuries, the young lacked such signals, it is questionable whether the species would have survived.

Bowlby's work, of course, has its limitations. For one thing, it focuses fairly exclusively on infancy. We would like ethological insights into later periods of development. To an extent, other ethologists have begun to fill this gap. For example, Blurton Jones (1972) and Daniel Freedman (1974) have studied dominance hierarchies and aggressive behavior in older children. We also expect that ethologists will have interesting things to say about adolescent behavior, when the urge toward independence and sexual behavior must bear analogies to behavior in other species. And, hopefully, ethologists will provide insights into family life and other behavior in the adult years.

The major drawback to Bowlby's work, I think, is his casual attitude toward certain conceptual matters. For example, he says that smiling, sucking, grasping, and following can all loosely be considered "instinctive" behaviors, but some of these might come closer to a rigorous definition of instinct than others. When a one-year-old urgently follows his or her mother, the behavior seems to be elicited by a rather specific stimulus—the mother's departure beyond a certain point. As such, the following response meets a fundamental require-ment of a strict instinct concept. Similarly, the baby's social smile also seems to have a specific releasing stimulus—the other's face in a frontal position. However, a behavior such as crying is prompted by numerous stimuli, both internal and external; the specific releasing stimulus is less clear. Perhaps crying is better thought of as a simple reflex. Thus, a more exacting analysis of the various attachment behaviors would seem to be in order.

These criticisms notwithstanding, Bowlby and the ethologists are making us realize the importance of viewing human development in an evolutionary context. They also are inspiring a new humility when we consider children. They are making us realize that we cannot change children's behavior in any way we wish. Evolution has provided children with responses and signals that must be heeded if development is to proceed properly. If, for example, we do not attend to children's cries, smiles, and other signals, they do not form the secure attach-ments which seem so necessary for later social growth. Before we try to change children's behavior, then, we should first understand how it follows Nature's plan for healthy development. By asking for a respectful attitude toward Nature, the ethologists are very much in the developmental tradition.

4

Montessori's Educational Philosophy

BIOGRAPHICAL INTRODUCTION

Most of the developmentalists whom we discuss in this book had ideas on education, but only Maria Montessori dedicated herself to the actual teaching of children. Montessori (1870-1952) was born in the province of Ancona, Italy. Her father was a successful civil servant with traditional ideas on the role of women in society. Her mother, in contrast, hoped that Montessori would go as far as she could in life. It was this hope that took hold. It is said that when Montessori was seriously ill as a 10-year-old, she told her anxious mother, "Don't worry, Mother, I cannot die; I have too much to do" (Kramer, 1976, p. 28). At the age of 26, Montessori became the first woman physician in Italy's history.

Montessori's first professional interest was in mental retardation. She was impressed by the extent to which institutionalized retarded children hungered for experience; she felt that they might be teachable if the right methods were used. She read as much as she could find on the subject and found that her own intuitions had guided an earlier line of educators, including Itard, Seguin, and Froebel—men who had worked in the spirit of Rousseau. According to these men, we cannot simply begin teaching retarded children things that we think they ought to know, such as reading and writing. This will only lead to frustration; for these children are not intellectually ready to learn on this level. Instead, we must first simply observe the children and take note of their natural tendencies and spontaneous interests. Then we will be in a position to take advantage of their own natural inclinations and ways of learning. For example, Seguin found that retarded children, like normal children at younger ages, are most interested in objects that stimulate their senses and permit physical activity. Accordingly, he gave them objects to place in different sized holes, beads to thread, pieces of cloth to button and lace, and other concrete and useful tasks (Kramer, 1976, p. 61).

Montessori followed Seguin's approach, using many of his materials and trying out new materials of her own. To her delight, this new approach worked, and she ventured to teach more difficult matters, including reading and writing, in the same way. Since the retarded children seemed to learn best by touching and feeling objects, she gave them wooden script letters that they liked to run their hands over again and again. By such methods, she taught many of these children to read and write as well as normal children of their age.

During her work with the retarded children, Montessori worked closely with another physician, Dr. Montessano, with whom she had a love affair. The result was a son, Mario. Montessori and Montessano never married, apparently because his parents objected (Kramer, 1976, p. 92). At that time in Italy, news of an illegitimate child would have ruined her career, so she followed the advice of her friends and secretly sent Mario to a wet nurse in the country. She did continue to visit her son, who later became an important educator in the Montessori movement. Nevertheless, the episode threw Montessori into a crisis, which she weathered by deepening her Catholic faith.

In 1907 Montessori took over responsibility for educating children who lived in a tenement in the slums of San Lorenzo, a section of Rome. There she established a school for over 50 extremely poor children—the sons and daughters of unemployed laborers, beggars, prostitutes, and criminals. In this school—called the Casa dei Bambini, or Children's House—Montessori continued to develop her ideas and techniques, and she was so successful that by 1913 she was one of the most famous women in the news. It seemed that her ideas were about to change the course of education throughout the world. However, her ideas apparently turned out to be too radical for the educational mainstream, and within five years she was all but forgotten except by a small band of followers. It was not until the last decade or so that her work once again began to catch the attention of psychologists, educators, and the general public (Lillard, 1972; Kramer, 1976).

THEORY OF DEVELOPMENT

Although Montessori's interests were more practical than theoretical, she did develop a definite theoretical position, one that owed much to Rousseau. She argued that we are wrong to assume that children are whatever we make them, for children also learn on their own, from their own maturational promptings (Montessori, 1936a, p. 22; 1949, p. 17, 223). And, like Rousseau, she argued that children often think and learn quite differently from adults (Montessori, 1936b, p. 69).

A central component of Montessori's theory is the concept of *sensitive periods*. Sensitive periods are similar to critical periods; they are genetically programmed blocks of time during which the child is especially eager and able to master certain tasks. For example, there are sensitive periods for the acquisition of language and for the beginning use of the hand. During these periods, the child works with all his or her might at perfecting these abilities. And, "if the child is prevented from enjoying these experiences at the very time when nature has planned for him to do so, the special sensitivity which draws him to them will vanish, with a disturbing effect on development . . ." (Montessori, 1949, p. 95).

The Sensitive Period for Order

During the first sensitive period, which takes place primarily during the first three years, the child has a strong need for order.[1] As soon as children can move about, they like to put objects where they belong; if a book or a pen is out of place, they resolutely put it back. And even before this, they often become upset at the sight of something out of order. Montessori told, for example, of a six-month-old girl who cried when a visitor put an umbrella on the table. The girl looked at the table and cried for some time. She became calm only when the mother, with a flash of insight, put the umbrella on the rack where it belonged (Montessori, 1936b, p. 50).

To us, such reactions, which are quite common, seem silly. This is because the adult need for order is on a different plane. For the adult, order provides a certain measure of external pleasure, but for the young child, it is essential. "It is like the land upon which animals walk or the water in which fish swim. In their first year [infants] derive their principles of orientation from their environment which they must later master" (1936b, p. 53).

The Sensitive Period for Details

Between one and two years of age, children fix their attention on minute details. For example, they detect small insects which escape our notice. Or, if we show them pictures, they seem to disregard the main objects, which we consider important, and focus instead on tiny objects in the background. This concern for details signals a change in children's psychic development. Whereas

[1] Montessori was rather vague about the ages of her sensitive periods, so the ages listed in this chapter are not definite.

they were at first attracted to gaudy objects and brilliant lights and colors, they are now trying to fill in their experience as completely as possible. To adults, the small child's concern for the minutest details is perplexing. It is further evidence that a child's "psychic personality is far different from our own, and it is different in kind and not simply in degree" (1936b, p. 69).

The Sensitive Period for the Use of Hands

A third sensitive period involves the use of the hands. Between about 18 months and three years of age, children are constantly grasping objects. They particularly enjoy opening and shutting things, putting objects into containers and pouring them out, and piling objects up (1936b, p. 83). During the next two years or so, they refine their movements and their sense of touch. For example, four-year-olds enjoy identifying objects by touching them with their eyes closed— a game which has far greater interest for the child than the adult (Montessori, 1948, p. 127, 229).

The Sensitive Period for Walking

The most readily visible sensitive period is for walking. Learning to walk, Montessori said, is a kind of second birth; the child passes from a helpless to an active being (Montessori, 1936b, p. 77). Children are driven by an irresistible impulse in their attempts to walk and they walk about with the greatest pride as they learn how.

We frequently fail to realize that walking, like other behaviors, means something different to the child than it does to us. When we walk, we have a destination in mind; we wish to get somewhere. The toddler, in contrast, walks for the sake of walking. For example, the child may walk up and down the staircase, over and over. The child does not walk to get somewhere, but to "perfect his own functions, and consequently his goal in something creative within himself" (1936b, p. 78).

The Sensitive Period for Language

A fifth sensitive period—and perhaps the most remarkable one of all— involves the acquisition of language. What is remarkable is the speed with which children learn such a complex process. To learn a language, children must learn not just words and their meanings, but a grammar, a system of rules which tells them where to place the various parts of speech. If, for example, we say, "The tumbler is on the table," the meaning we give those words derives from the order in which we say them. If we had said, "On tumbler the is table the," our meaning would have been hard to grasp (Montessori, 1949, p. 25). The rules underlying grammars are so elusive and abstract that linguistic scholars are still trying to understand them in a formal way. Yet children master them without much thinking about it. If a child is exposed to two languages, the child masters them both (1949, p. 111).

Because the child's ability to grasp language is so great, Montessori concluded, the child must be endowed with a special kind of language receptivity or "mechanism" (1949, p. 113). This mechanism is very different from anything in the mental life of the older child or the adult. Whereas we learn a second language with great deliberation, consciously struggling to memorize rules regarding tenses, prefixes, modifiers, and so on, the child absorbs language *unconsciously*.

From Montessori's descriptions, the child's language acquisition sounds very much like a kind of imprinting. At a certain critical time—from the first few months of life until two and a half or three years—children are innately prepared to absorb sounds, words, and grammar from the environment. "The child *absorbs* these impressions not with his mind, but with his life itself" (1949, p. 24). Sounds create impressions of incredible intensity and emotion; they must set in motion invisible fibers in the child's body, fibers which start vibrating in the effort to reproduce those sounds (1949, p. 24). We adults can hardly imagine what this experience is like, except perhaps by recalling the feeling we get when we are profoundly moved by a symphony, and then imagining a similar feeling which is several times stronger. This particular sensitivity for language comes into play during the first three years or so, and then is gone.

Montessori suggested that, because language acquisition is governed by innate, maturational factors, children develop language in the same stages no matter where they grow up (1949, p. 111). For example, they all proceed from a stage of babbling to a stage where they begin speaking words. Next, they enter a stage in which they put two-word sentences together (e.g., "Boy go"), and there follows a period in which they master increasingly complex sentence structures.

These stages, Montessori emphasized, do not emerge in a gradual, continuous manner. Instead, there are several times during which the child

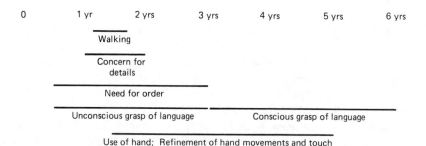

FIGURE 4.1   Some early sensitive periods.

seems to be making no progress, and then new achievements come in explosions. For example, the child bursts out with a number of new words, or suddenly masters a set of rules for forming parts of speech, such as suffixes and prefixes, in a sudden explosion (1949, p. 114).

Between the ages of about three and six years, children no longer absorb words and grammar unconsciously, but they still are in the general sensitive period for language. During this time, they are more conscious of learning new grammatical forms and take great delight in doing so.

By the time the child is five or six, then, and is ready for the traditional school, the child has already learned to talk. "And all this happens without a teacher. It is a spontaneous acquisition. And we, after he has done all this by himself, send him to school and offer as a great treat, to teach him the alphabet!" (1949, p. 115). We are also so presumptuous as to focus on the child's defects, which are trivial in comparison to the child's remarkable achievements. We see how adults assume that what they teach the child amounts to everything and how they grossly underestimate the child's capacity to learn on his or her own.

EARLY EDUCATION IN THE HOME

At various sensitive periods, then, children are driven by an inner impulse to independently master certain experiences. The goal of education is to assist this process. Since children do not ordinarily enter nursery or Montessori schools until they are two or three years old, parents and caretakers are really the first educators.

To be of help, we do not necessarily need a formal knowledge of child psychology. What we need is a certain attitude. We need to realize that it is not our job to direct our children's learning, for we must, above all, respect their efforts at independent mastery. We must have a faith in their powers of inner construction. At the same time, we do not have to simply ignore or neglect our children. What we can do is give them the opportunities to learn what is most vital to them. We can watch for their spontaneous interests and give them the opportunities to pursue them.

For example, Montessori (1936b, p. 49) told about a nurse who pushed her five-month-old girl in the carriage through the garden. Each day, the girl was delighted to see a white marble stone cemented to an old gray wall. Apparently the girl took pleasure in her discovery that the stone was in the same place; she was developing her sense of order in the world. In any case, the nurse, noting the girl's interest, stopped each day to permit the child to look at the sight. The nurse did not just push the girl along at her own pace, but let the infant's spontaneous interest guide her. She was not teaching the child in the ordinary sense, but she was behaving like the ideal teacher. She was giving the baby the opportunity to make whatever spontaneous connection she was working on.

To take another example, parents can help their children during the sensitive period for walking. Some parents enjoy following their child about as he or she masters this new skill. They follow their children as they travel amazingly long distances, and stop with their children when they stop to examine new sights. They also give their children time to master new aspects of walking, such as stairclimbing. They follow their children's own rhythms and enjoy watching the pride children take in mastering this new skill (1936b, Ch. 11).

Other parents, however, fail to give their children full opportunities to master walking on their own and in their own way. Some try to teach the child to walk—a practice which probably gives the child the sense that his or her own efforts are inadequate. Or parents fail to realize what walking means to the child; they assume that the child, like them, wants to get somewhere. So they pick up their children and carry them, or push them in strollers, so as to reach their destination more rapidly. Or parents become afraid of where the child's walking may lead, so they surround the child with safeguards, such as playpens (1936b, Ch. 11).

All normal children, of course, do eventually learn to walk, but the parental reactions affect the children's feelings about their inner rhythms and independence. In the one case, they gain the feelings of freedom and pride that come from mastering an important skill on one's own. In the other case, they find that their own efforts at mastery produce negative reactions. The result may be a lingering inhibition with respect to one's inner promptings. It is possible that later athletic ability and physical grace are related to the manner in which children first learn to walk.

THE MONTESSORI SCHOOL

When children are two or two and a half years old, they can enter a Montessori school. There they learn in the same class with children up to six or seven years of age—approximately the same age range as in Montessori's Children's House. In some schools, children may next join classes with those up to age 11 or so. In any case, the ages are still mixed, because Montessori found that children enjoy this arrangement.

Independence and Concentration

The goal of education in the school is the same as that in the home. The teacher does not try to direct, instruct, drill, or otherwise take charge of the child; instead, the teacher tries to give the child opportunities for *independent* mastery. The assumption is that if the school environment contains the right materials—those that correspond to the children's inner needs at various sensitive periods—the children will enthusiastically work on them on their own, without adult supervision.

To create the right environment, Montessori first spent considerable time simply observing children's behavior with respect to various materials. She then retained those that the children themselves told her were the most valuable to them. They told her this in a rather dramatic way; when they came across materials that met deep, inner needs, they worked on them with amazing *concentration.*

Montessori first became aware of the child's capacity for concentrated effort when she saw a four-year-old girl in the Children's House working on cylinders. That is, the girl was placing different sized cylinders in the holes of a wooden frame until she had them all in place (see Fig. 4.2). Then she would take them out again, mix them up, and start over. All the while she was oblivious to the world around her. After she had repeated this exercise 14 times, Montessori decided to test her concentration. She had the rest of the class sing and march loudly, but the girl simply continued with her work. Montessori then

FIGURE 4.2   A boy works on wooden cylinders. (St. Michael's Montessori School, New York City. Haledjian photo.)

lifted the girl's chair—with the child in it—onto a table. But the girl merely gathered up her cylinders in her lap and kept on working, hardly aware of the disturbance. Finally, after 42 repetitions, the child stopped on her own, as if coming out of a dream, and smiled happily (Montessori, 1936b, p. 119).

After that, Montessori observed the same phenomenon on countless occasions. When given tasks that met inner needs at sensitive periods, the children worked on them over and over. And when they finished, they were rested and joyful; they seemed to possess an inner peace. It seemed that children were achieving, through intense work, their true or normal state; Montessori therefore called this process *normalization* (1949, p. 206). She then made it her goal to create the most favorable environment for this kind of concerted effort.

Free Choice

In preparing her environment, Montessori tried to suspend her own ideas about what children should learn and to see what they selected when given a free choice. Their free choices, she learned, usually led to work on the tasks that most deeply engrossed them. For example, she noticed that the two-year-olds, when free to move around in the room, were constantly straightening things up and putting them in order. If, for example, a glass of water slipped from a child's hands, the others would run up to collect the broken pieces and wipe the floor (Montessori, 1936b, p. 121). What she observed was their need for order. Accordingly, she altered the environment so they could fulfill this need. She made small washbasins so the children could wash their hands and brushes; she made small tables and chairs, so they could arrange them just right; and she lowered the cupboards, so they could put their materials away where they belonged. In this way, activities of daily living became part of the curriculum. All the children enjoyed these activities, but the two-year-olds took them the most seriously. They constantly inspected the room to see if anything were out of place. For them, ordering the environment met the deepest inner need (Montessori, 1948, p. 88).

Today, the core Montessori materials are largely set, but the teacher still relies heavily on the principle of free choice. Each child goes to the cupboard and selects the apparatus that he or she wants to work on. The teacher has a faith that the children will freely choose the tasks that meet their inner needs at the moment.

Although the teacher permits free choice, the teacher will from time to time introduce a child to a new task that he or she seems ready for. This is done most delicately. The teacher presents the material clearly and simply and then steps back to observe the child's behavior. The teacher watches for concentration and repetition. If the child does not seem ready for the new task, it is put aside for another day. The teacher must be careful to avoid giving the impression that the child "ought" to learn a particular task; for this would undermine the child's ability to follow his or her tendencies. If the child does begin to work actively on the material, the teacher moves away and lets the child work independently (Lillard, 1972, pp. 65-68).

The teacher's attitude, Montessori said, is essentially a passive one—that of an observer (Montessori, 1936a, p. 39). He or she spends most of the time simply watching the children, trying to guess each one's particular needs and state of readiness.

Rewards and Punishments

The Montessori teacher, then, is not so much a director, but a follower. It is the child who leads the way, revealing what he or she most needs to work on. In this, the Montessori teacher behaves very differently from the typical teacher, who has set goals for the children and tries to take charge of their education.

The typical teacher often finds that children lack enthusiasm for the things they are asked to learn. The teacher therefore relies heavily on external rewards and punishments—praise, grades, threats, and criticism. Yet these external inducements often seem to backfire. Quite often children become so concerned with external evaluations—so afraid of getting wrong answers and looking stupid—that they cannot concentrate deeply on their work. Driven by such pressure, they will learn a certain amount of material, but they can easily come to dislike school and the learning process (Montessori, 1948, p. 14; Holt, 1964).

Equally damaging, external evaluations rob children of their independence. Children soon begin looking to external authorities, such as the teacher, to know what they should do and say. Montessori felt that authorities use rewards and punishments primarily to make the child submissive to their will. Like Rousseau, she wondered how a child who becomes anxious about external approval will ever learn to think independently or will ever dare to criticize the conventional social order (Montessori, 1948, pp. 14-18).

Thus, rewards and punishments have no place in the Montessori classroom. Montessori teachers trust that if they pay attention to children's spontaneous tendencies, they can find the materials on which children will work intently on their own. The children will do so out of an inherent drive to perfect their capacities, and external inducements will become superfluous.

The traditional teacher often justifies reward and criticism as necessary because children need to know when they are right or wrong. Montessori agreed that children need to learn from their errors, but she did not want the children to have to turn to adults for this information. Accordingly, she developed many materials with a built-in *control of error*. For example, the cylinders, which teach spatial dimensions, have this control. If a child has not put each cylinder in the proper hole, there will be one cylinder left over. When children see this, their interest is heightened. They then figure out how to correct the matter on their own. Thus, control of error enables them to correct their own mistakes, without relying on adults for criticism or advice.

Gradual Preparation

Montessori found that children cannot learn many skills all at once. For example, four-year-olds often desperately want to learn to button their coats and tie their own shoes, as a consequence of their natural urge toward independence, but these tasks are too difficult for them. They lack the fine motor skills.

To deal with such problems, Montessori developed materials that would enable the children to learn skills in steps, at levels they could master. In the case of shoe-tying, she developed a large tying frame (Figure 4.3) so the children could practice the correct typing pattern with grosser muscle movements (1948, p. 93). She also utilized the principle of indirect preparation (1948, p. 224). That is, she gave them unrelated tasks, such as cutting vegetables (Figure 4.3) and holding a pencil, through which they could simultaneously perfect their dexterity. Then, when the children decided to attempt to tie their own shoes, they could readily do so, for they had gradually mastered all the necessary subskills.

Reading and Writing

I have mentioned, by way of illustration, some of the tasks that are part of the Montessori method (e.g., the cylinders and the exercises of daily living). We cannot review every component of the Montessori curriculum in this book, but I will try to indicate how Montessori approached one important area—reading and writing.[2]

Montessori found that if one begins at the age of about four years, children will learn to read and write with great enthusiasm. This is because they are still in the general sensitive period for language. They have just mastered language unconsciously and are now eager to learn all about it on a more conscious level, which reading and writing permit them to do. If, in contrast, one waits until the age of six or seven years to teach written language, as schools usually do, the task is more difficult because the sensitive period for language has already passed (1948, p. 276).

Four-year-olds will usually master writing before reading. This is because writing is the more concrete and sensory activity and therefore better suits the young child's style of learning (1948, p. 233). Still, one cannot teach writing all at once. If one asks four-year-olds to make a sound and write it, they will be unable to do so; one must introduce writing through a series of separate preparatory exercises.

First, the child is shown how to hold a pencil and then practices drawing by staying within outlines. Children love to practice drawing as precisely as

[2] For a summary of the method for teaching arithmetic, see Montessori, 1948, Chs. 18 and 19.

FIGURE 4.3   Children at work on a tying frame and an exercise of daily living. (St. Michael's Montessori School, New York City. Haledjian photo.)

possible, for they are in the sensitive period for precise hand movements. They also have been mastering precise hand-eye coordination through exercises of daily living, such as cutting vegetables, pouring, and polishing silver.

In another exercise, children trace their fingers over sandpaper letters which are pasted onto blocks of wood (see Figure 4.4). For example, they make the "m" sound and trace it as they do so. The letters are written in script, rather than print, because children find the movements of script freer and more natural. Through this exercise, then, they learn to make the movements of the letters. They love repeating this exercise, for they are still in the sensitive periods for learning about sounds and refining their sense of touch. Frequently they like to close their eyes and trace the letters with their fingers alone. Six-year-olds, in contrast, derive no particular pleasure from the sandpaper letters, for they have already moved out of the sensitive period for touch. The letters, incidentally, have a built-in control of error, since the child can tell when his or her finger has strayed off the letter and onto the wood because the wood feels different (1948, p. 229).

FIGURE 4.4   A girl works on sandpaper letters. (St. Michael's Montessori School, New York City. Haledjian photo.)

In a third exercise, children are given a moveable alphabet which permits them to form the letters of words. For example, they look at a picture of a cat, sound out the letters, and then make the word with the letters. This, too, they repeat endlessly, out of their spontaneous interest in the elements of spoken language (1948, pp. 234-37).

Through these and other separate exercises, then, children learn the various skills involved in writing. When they finally put these skills together and begin to write letters, there usually follows an "explosion of writing." They will write all day long (1948, p. 239).

Writing paves the way for reading. Through writing, children form a muscular and visual memory of the letters and words and therefore can recognize them. Consequently, the five- or six-year-old who has learned to write can usually learn to read with very little help from the teacher (Lillard, 1972, p. 122). Children often say that nobody taught them to read at all. Montessori did aid the process, though. Her essential method was to show a word printed on a card, ask the child to sound it out, and then ask the child to sound it out more quickly. In most cases, children rapidly catch on and begin reading words on their own.

During the entire preparatory period for writing and reading, the children do not even look at a book. Then, when they first pick a book up, they usually can begin reading it immediately. Consequently, they avoid all the frustrating experiences that children so often associate with books. There follows an "explosion of reading." Children delight in reading everything they see (Montessori, 1948, p. 253).

The sensitive care with which Montessori prepared each small step is impressive. The exercises are arranged so that each comes easily to the child, for each corresponds to the child's natural way of learning. The method contrasts sharply with that of most teachers, who simply give children lessons and then spend most of their time criticizing them for their mistakes. Criticism, Montessori felt, is humiliating and pointless. Instead of criticizing, which only tears down, the teacher should figure out ways of helping children build their skills (1949, p. 245).

Misbehavior

We have emphasized how Montessori teachers prize the child's independence—how they avoid imposing expectations on the child, or even praising or criticizing the child. This is true with respect to *intellectual* work. *Moral* misconduct is another matter. Children are not permitted to abuse the materials or their classmates.

In the Montessori school, respect for the materials and for others usually develops quite naturally. The children know how important the work is for themselves, so they respect the work of the other children. If they do bother a child who is in deep concentration, the child usually insists on being left alone in such a way that they automatically respect this wish. Sometimes, though, the teacher must intervene. Montessori (1948, p. 62) recommended isolating the child for a moment. In this way, the child has a chance to see the value of the work for others and to sense what he or she is missing. The child will then begin constructive work without any further prompting.

In general, the Montessori view of discipline is different from that of most teachers. Most teachers think that it is their job to gain control over the class. They shout: "I want everyone in their seats!" "Didn't you hear what I just told you?" "If you don't behave this instant you won't go out for recess!" The Montessori teacher is not interested in such obedience. Real discipline is not something imposed from without, from threats or rewards, but something that comes from within, from the children themselves as they "pass from their first disordered movements to those that are spontaneously regulated" (1948, p. 56).

Misbehavior, in Montessori's view, usually indicates that the children are unfulfilled in their work. Accordingly, one's task is not to impose one's authority on the children but to observe each child more closely, so one will be in a better position to introduce materials that will meet his or her inner

---

BOX 4-1   Two Six-year-old Boys' Views on School Matters

Note the differences in the role of the teacher in the minds of these two children.

1. Who taught you to read?

   Regular School Child: "My teacher."

   Montessori Child: "Nobody. I just read the book, and to see if I could read it."

2. Do you get to work on anything you want?

   Regular School Child: "No. But we can go to the bathroom anytime we want. But we're not allowed to go to the bathroom more than four times."

   Montessori Child: "You can work on anything you want."

3. What would happen if you bothered another kid who was working?

   Regular School Child: "I'd get in trouble from the teacher."

   Montessori Child: "He'll just say, 'Please go away, I'm busy.'" (What would you do?) "I'd just go away, 'cause I don't want to bother someone working."

developmental needs. The teacher expects a certain amount of restlessness and distracted behavior during the first days of the year, but once the children settle into their work they become so absorbed in it that discipline is rarely a problem.

Fantasy

Montessori was critical of attempts to enrich children's fantasy lives through fairy tales, fables, and other fanciful stories. She saw fantasy as the product of a mind that has lost its tie to reality (Montessori, 1917, p. 255). The insane lose their ability to deal with the real world and therefore turn to inner imaginings.

Montessori's position on fantasy would appear to contradict one of her most basic tenets: that we should follow children's natural inclinations. For children, she acknowledged, do have a natural bent toward fantasy. As she put it, the child's "mentality differs from ours; he escapes from our strongly marked and restricted limits, and loves to wander in the fascinating worlds of unreality..." (1917, p. 255). But our task is to help the child overcome these tendencies. When we read children fairy tales or tell them about Santa Claus, we only encourage their credulity. When they hear these stories, furthermore, their basic attitude is passive; they simply take in the impressions we give them. They believe fantastic things because they have not yet developed their powers of discrimination and judgment. And it is just these powers that they need to build.

Montessori did recognize the uses of a creative imagination, such as that possessed by the artist. But the artist's creativity, she maintained, is always tied to reality. In fact, the artist sees things more clearly than we do. The artist is more sensitive to forms, colors, harmonies, and contrasts. And "it is by refining his powers of observation that the artist perfects himself and creates a masterpiece" (1917, pp. 250-51). If we wish to help children become creative, then, we need to help them develop their powers of observation and discrimination with respect to the real world, and not encourage them to drift off into an unreal one.

Summary

We ordinarily assume that it is our task to mold or shape our children, to teach them to learn and to think, but we are mistaken. Children learn on their own, from their inner need to develop their capacities. We assist children's development by preparing the right environment, an environment which contains materials that correspond to their inner needs at various sensitive periods. For example, since young children have a need for order, furniture and other objects should be built in such a way that they can arrange them. In this way, they can actively develop their sense of order and orientation in the world. To a considerable extent, children can be trusted to choose freely the materials that they most need, and they will work on them with the greatest concentration. From time to time, though, teachers can introduce new tasks, based on their guesses about the individual child's sensitive period. For example, when a child shows an interest in written language, the teacher can introduce an appropriate preparatory task, such as the sandpaper letters. The teacher then watches for concentration and repetition. If the sandpaper letters do in fact correspond to the child's sensitive period, the child will become deeply absorbed in tracing them. In this situation, praise and criticism are unnecessary. Worse than this, praise and criticism undermine concentration and independence by creating a concern for external evaluation. It is far better for children to correct their own mistakes, which the proper materials enable them to do.

Although the Montessori teacher would never criticize, let alone punish a child for an intellectual error, certain misbehavior is not tolerated. Children must respect the materials and the work of others. Misbehavior, however, usually indicates that a child is unfulfilled in his or her work. Instead of commanding obedience, the teacher looks for ways to get the child absorbed in work. The Montessori school, finally, does not encourage fantasy. Children develop themselves by mastering the real world.

EVALUATION

Little systematic research has been done on the effectiveness of Montessori education (see Lillard, 1972, pp. 155-59), but those who visit a Montessori school usually come away impressed. One is struck by the quiet dignity of the classroom; the atmosphere is almost that of a monastery, with everyone seriously at work. There is no teacher shouting, and the children show respect for one another.

Meaningful evaluations of the Montessori method should be mindful of Montessori's objectives. Her primary goal was not high scores on achievement tests. There is no reason to suppose that Montessori children lag behind here; we have seen, for example, that they usually learn to read and write at an early age. But in terms of overall philosophy, this is just a fortunate happenstance. Montessori chose to teach writing to four-year-olds only because they revealed an inner urge to write at this young age. If she had found no such urge until, say, age 10, she would not have taught it until then. She did not want to impose tasks on children just because adults are anxious that they learn them as soon as possible. She cared little about how rapidly children learn standard skills or about advancing them along the ladder of achievement tests. Rather, her goals had to do with inner attitudes. She wanted to unharness the child's capacity for concerted and independent work, a capacity which unfolds according to an inner timing. As she once said,

> My vision of the future is no longer of people taking exams and proceeding on that certificate from the secondary school to the University, but of individuals passing from one stage of independence to a higher, by means of their own activity, through their own effort of will, which constitutes the inner evolution of the individual (1936, cited in Montessori, 1970, p. 42).

Thus, meaningful evaluations should examine not only test performances but children's "intellectual character"—their independence, self-composure, love of learning, and tendency to lose themselves in their work. On such variables, most parents I know would rate their Montessori school children quite high indeed.

Although Montessori is well known as a teacher, she is underestimated as an innovative theoretician. She anticipated much that is current in developmental thinking. For one thing, she was among the first to argue for the possibility of sensitive or critical periods in intellectual development. Even more impressive were her insights into language acquisition. Early on, she suggested that chidren unconsciously master complex grammatical rules and suggested that they must possess an innate mechanism which enables them to do this—suggestions which anticipated the work of Chomsky (see Chapter 14 in this book).

What are the criticisms of Montessori? John Dewey, who shared many of Montessori's goals, thought her methods might limit creativity. If children are not ready for an apparatus, they are not permitted to merely play with it, and the materials lend themselves only to limited uses (Dewey and Dewey, 1915, p. 157). Montessori's followers counter that at the right moment the materials call forth the child's deepest creative urges. Children become so absorbed in them precisely because they are thrilled by their discoveries (Lillard, 1972, p. 69).

A more fundamental criticism is that Montessori emphasized cognitive development to the exclusion of the child's social and emotional life. In the Montessori classroom, the children work primarily with physical objects and do not interact much either with other children or the teacher. The teacher introduces the material and then fades into the background, allowing the child to work on his or her own. In the Montessori view, things, not people, are the best teachers (Kramer, 1976, p. 21). This method fosters independence, but it also leaves a great deal out; ". . . it seems to leave out a whole dimension of being human—the emotional" (Kramer, 1976, p. 21). Montessori, of course, was justifiably suspicious of learning based on emotional involvements with others. It is all too easy for children to become trapped in the process of learning to please grown-ups rather than learning for its own sake. Nevertheless, many people believe that Montessori went too far in the opposite direction, making learning too impersonal.

Montessori's attitude toward the child's inner emotional life is exemplified by her view toward fantasy and fairy tales. She admitted that children possess a natural bent toward fantasy and enjoy fairy tales, but she discouraged them anyway. Imaginary stories only encourage children to depart from reality. Further, when children listen to fairy tales, they passively receive impressions from adults. They do not think on their own (Montessori, 1917, p. 259).

It is not clear, however, that Montessori was right. In a recent book (1976), Bettelheim points out that fairy tales do not really teach children to believe in imaginary happenings, for children know that the fairy tale is make-believe. The stories themselves make this point with their opening lines—"Once upon a time," "In days of yore and times and tides long ago," and so on (Bettelheim, 1976, p. 117). Children intuitively understand that the story addresses itself not to real, external events, but to the inner realm of secret hopes and anxieties. For example, "Hansel and Gretel" deals with the child's fear of separation and does so in a way that points to a solution. It indirectly encourages children to become independent and use their own intelligence.

Furthermore, the process of listening to a fairy tale may be much more active than Montessori realized. When children listen to a story, they interpret it in their own way and fill in the scenes with their own images.[3] When a story speaks to an issue with which the child is inwardly struggling, the child wants to hear it over and over, much as Montessori children work repeatedly on external exercises. And, finally, children often emerge from the story in a state of calm and peace, as if they have resolved some issue (Bettelheim, 1976, p. 18).

Montessori, then, may have been too quick to dismiss the child's inner world of fantasy and the value of fairy tales. Nevertheless, such oversights seem minor in comparison to her contributions. Montessori demonstrated, as much or more than anyone else, how the developmental philosophies of Rousseau, Gesell, Piaget, and others can be put into effective practice. She showed how it is possible to follow children's spontaneous tendencies and to provide materials that will permit them to learn on their own. Montessori must be regarded as one of history's greatest educators.

---

[3] The process of listening to a fairy tale seems much more active, for example, than most television watching. Television itself usually supplies the child with all the images (Singer and Singer, 1979).

5

Piaget's Cognitive-Developmental Theory

BIOGRAPHICAL INTRODUCTION

Among contemporary psychologists, there is no greater theorist than Jean Piaget. Almost single-handedly, he has forged a comprehensive and compelling theory of cognitive development—a theory of the growth of the intellect.

Piaget was born in 1896 in Neuchâtel, a small college town in Switzerland where his father was a medieval historian at the University. Piaget (1952) describes his father as a careful and systematic thinker. His mother, in contrast, was highly emotional, and her behavior created tensions within the family. Piaget adopted his father's studious ways and found refuge from the family's conflicts in solitary research.

Piaget showed promise as a scientist from the start. At the age of 10, he published an article on an albino sparrow he had seen in the park. While still in high school, his research on mollusks brought invitations to meet with foreign colleagues and a job offer to become the curator of a museum—all of which he turned down because of his age.

At 15, Piaget experienced an intellectual crisis when he realized that his religious and philosophical convictions lacked a scientific foundation. He therefore set out to find some way of bridging philosophy with science. He read widely and worked out new ideas in writing, even though the writing was intended for no one but himself. This search did not occupy all his time—he still managed to earn his doctorate in the natural sciences at the age of 21—but Piaget's broader quest did at times leave him confused and exhausted. Finally, at the age of 23, he settled on a plan. He would first do scientific research in child psychology, studying the development of the mind. He then would use his findings to answer broader questions in epistemology, philosophical questions concerning the origin of knowledge. He called this new enterprise "genetic epistemology" (Piaget, 1952, pp. 239-44; Ginsburg and Opper, 1969, pp. 2-3).

Piaget decided to study children in 1920 while working in the Binet Laboratory in Paris. There his assignment was to construct an intelligence test for children. At first he found this work very boring; he had little interest in scoring children's answers right and wrong, as intelligence testing requires. However, Piaget soon became interested in the younger children's responses, especially their *wrong* answers. Their mistakes, he found, fit a consistent pattern that suggested that their thinking might have a character all its own. Young children, Piaget speculated, might not simply be "dumber" than older children or adults, but might think in an entirely different way (Ginsburg and Opper, 1969, p. 3).

In order to learn about children's potentially unique ideas, Piaget abandoned the standardized tests, which forced children's responses into "artificial channels of set question and answer," and devised a more open-ended clinical interview which "encourages the flow of spontaneous tendencies" (Piaget, 1926, p. 4). He also spent many hours observing children's spontaneous activities. The point was to suspend his own adult preconceptions about children's thinking and to learn from the children themselves.

While in Paris, Piaget published two studies based on his new approach, but he did most of this new research at the Rousseau Institute in Geneva, where he has been since 1921. He primarily interviewed children between the ages of four and 12 years, and he found that the younger children, before the age of seven or so, do indeed think in a qualitatively different way about dreams, morals, and many other topics.

In 1925, Piaget's first child, Jacqueline, was born—an event which initiated an important series of studies on the cognitive behavior of infants. Piaget and his wife, Valentine Châtenay, made very careful observations of Jacqueline's behavior, as they also did of their next two babies, Lucienne and Laurent.

Beginning about 1940, Piaget returned to the study of children, and adolescents as well, but he changed his research focus. Whereas his earlier investigations covered such topics as dreams, morality, and other matters of everyday interest to the child, his new studies focused on the child's understanding of mathematical and scientific concepts—a focus which has dominated his work to the present (Ginsburg and Opper, 1969, p. 16).

In the 1950s, Piaget finally turned to philosophical questions in epistemology, although he has continued to study children's cognitive development. In this book, we will say little about Piaget's epistemological theory; our task will be to gain some understanding of his developmental theory.

Piaget's research has evoked different responses from psychologists at different times. His first work caught the attention of psychologists in many parts of the world. After this initial enthusiasm, however, interest in Piaget declined, especially in the United States. For one thing, psychologists had difficulty understanding his orientation. They also objected to his methodology. Piaget sometimes changed his questions during an interview if he thought this might help him understand a particular child's thinking; this, many psychologists pointed out, violates the canon of standardized interviewing. Piaget also ignored such matters as reports on his sample sizes and statistical summaries of his results. He seemed to regard such matters as less important than rich, detailed examples of children's thinking (Flavell, 1963, pp. 10-11, 431; Ginsburg and Opper, 1969, p. 7).

By and large, Piaget's research has suffered from the same methodological shortcomings throughout his career, but the 1960s saw a remarkable revival of interest in his work. Psychologists began to realize that his theory, no matter how difficult and how casually documented, is of tremendous importance. Today, there is hardly a study of children's thinking that does not refer to Piaget.

OVERVIEW OF THE THEORY

Although Piaget's research has changed over the years, each part of it contributes to a single, integrated stage theory. The most general stages, or periods, are listed in Table 5.1 below.

TABLE 5.1 The General Periods of Development

Period I.	Sensori-Motor Intelligence (birth to two years). Babies organize their physical action schemes, such as sucking, grasping, and hitting, for dealing with the immediate world.
Period II.	Preoperational Thought (two to seven years). Children learn to think—to use symbols and internal images—but their thinking is unsystematic and illogical. It is very different from that of adults.
Period III.	Concrete Operations (seven to 11 years). Children develop the capacity to think systematically, but only when they can refer to concrete objects and activities.
Period IV.	Formal Operations (11 to adulthood). Young people develop the capacity to think systematically on a purely abstract and hypothetical plane.

Before we examine these stages in detail, it is important to note two theoretical points. First, Piaget recognizes that children pass through his stages at different rates, and he therefore attaches little importance to the ages associated with them. He does maintain, however, that children move through the stages in an *invariant sequence*—in the same order.

Second, as we discuss the stages, it is important to bear in mind Piaget's general view on the nature of *developmental change*. Because he proposes an invariant stage sequence, some scholars (e.g., Bandura and McDonald, 1963) have assumed that he is a maturationist. He is not. Maturationists believe that stage sequences are wired into the genes and that stages unfold according to an inner timetable. Piaget, however, does not think that his stages are genetically determined. They simply represent increasingly comprehensive ways of thinking. Children are constantly exploring, manipulating, and trying to make sense out of the environment, and in this process they actively construct new and more elaborate structures for dealing with it (Kohlberg, 1968).

Piaget does make use of biological concepts, but only in a limited way. He observes that infants inherit reflexes, such as the sucking reflex. Reflexes are important in the first month of life but have much less bearing on development after this.

In addition, Piaget sometimes characterizes children's activities in terms of biological tendencies that are found in all organisms. These tendencies are assimilation, accommodation, and organization. *Assimilation* means taking in, as in eating or digestion. In the intellectual sphere, we have a need to assimilate objects or information into our cognitive structures. For example, adults assimilate information by reading books. Much earlier, a baby might try to assimilate an object by grasping it, trying to take it into his or her grasping scheme.

Some objects do not quite fit into existing structures, so we must make *accommodations,* or changes in our structures. For example, a baby girl might find that she can only grasp a block by first removing an obstacle. Through such accommodations, infants begin constructing increasingly efficient and elaborate means for dealing with the world.

The third tendency is *organization.* For example, a four-month-old boy might have the capacity to look at objects and to grasp them. Soon he will try to combine these two actions by grasping the same objects he looks at. On a more mental plane, we build theories. We seem to be constantly trying to organize our ideas into coherent systems.

Thus, even though Piaget does not believe that stages are wired into the genetic code, but constructed by children themselves, he does discuss the construction process in terms of biological tendencies (Ginsburg and Opper, 1969, pp. 17-19).

If Piaget is not a maturationist, he is even less a learning theorist. He does not believe that children's thinking is shaped by adult teachings or other environmental influences. Children must interact with the environment to develop, but it is they, not the external environment, who build new cognitive structures.

Development, then, is not governed by internal maturation or external teachings. It is an active construction process, in which children, through their own activities, build increasingly differentiated and comprehensive cognitive structures.

PERIOD I. SENSORI-MOTOR INTELLIGENCE (BIRTH TO TWO YEARS)

Piaget's first developmental period consists of six stages.

Stage 1 (birth to one month).[1]
The Use of Reflexes

When Piaget talks about the infant's action-structures, he uses the term *scheme* or *schema* (e.g., 1936a, p. 34). A scheme can be any action pattern for dealing with the environment, such as looking, grasping, hitting, or kicking. As mentioned, although infants construct their schemes and later structures through their own activities, their first schemes consist primarily of inborn reflexes. The most prominent reflex is the sucking reflex; babies automatically suck whenever their lips are touched.

Reflexes imply a certain passivity. The organism lies inactive until something comes along to stimulate it. Piaget, however, shows that even a reflex like sucking quickly becomes part of the human infant's self-initiated activity. For example, when his son Laurent was only two days old, he began making sucking movements when nothing released them. Since he did this between meals, when he wasn't hungry, he seemed to suck simply for the sake of sucking. Piaget says that once we have a scheme, we also have a need to put it to active use (1936a, pp. 25-26, 35).

Furthermore, when babies are hungry, they do not just passively wait for the mother to put the nipple into their mouth. When Laurent was three days old, he searched for the nipple as soon as his lips touched part of the breast. He groped, mouth open, across the breast until he found it (1936a, p. 26).

Babies do not confine themselves to sucking on nipples. Piaget's children sucked on clothes, pillows, blankets, their own fingers—on anything they chanced upon (1936a, p. 26, 34). In Piaget's terms, they assimilated all kinds of objects into the sucking scheme (1936a, p. 32).

Although assimilation is the most prominent activity during stage 1, we also can detect the beginnings of accommodation. For example, babies must learn to adjust their head and lip movements to find the breast and nurse. Such

[1] The age norms for this period are those suggested by Ginsburg and Opper (1969) in their excellent review of Piaget's theory.

adjustments also demonstrate the beginnings of organization; babies organize their movements so that nursing becomes increasingly smooth, rapid, and efficient (1936a, pp. 29-31, 39).

Stage 2 (one to four months). Primary Circular Reactions

A circular reaction occurs when the baby chances upon a new experience and tries to repeat it (1936a, p. 55). A prime example is thumb-sucking. By chance, the hand comes into contact with the mouth, and when the hand falls the baby tries to bring it back. For some time, however, babies cannot do this. They hit the face with the hand, but cannot catch it; or they fling their arms wildly; or they chase the hand with the mouth, but cannot catch it because the whole body, including the arms and hands, move as a unit in the same direction (1936a, pp. 51-53). In Piaget's language, they are unable to make the accommodations necessary to assimilate the hand to the sucking scheme. After repeated failures, they organize sucking and hand movements and master the art of thumb-sucking.

As with thumb-sucking, most of the primary circular reactions involve the organization of two previously separate body schemes or movements. For example, when we see a baby girl repeatedly bring her hand next to her face and look at it, she is exercising a primary circular reaction. She is coordinating looking with hand movements (1936a, pp. 96-97).

These circular reactions provide a good illustration of what Piaget means by intellectual development as a "construction process." The baby actively "puts together" different movements and schemes. It is important to emphasize the amount of work involved; the baby manages to coordinate separate movements only after repeated failures.

Stage 3 (four to 10 months). Secondary Circular Reactions

The developments of the second stage are called "primary" circular reactions because they involve the coordination of parts of the baby's own body. *Secondary* circular reactions occur when the baby discovers and reproduces an interesting event *outside* himself or herself (1936a, p. 154). For example, one day when Piaget's daughter Lucienne was lying in her bassinet, she made a movement with her legs which stirred the dolls hanging overhead. She stared at the dolls a moment and then moved her legs again, watching the dolls move again. In the next few days, she repeated this scene many times, kicking her legs and watching the dolls shake, and she often would squeal with laughter at the sight of the moving dolls (pp. 157-59).

Piaget sometimes refers to secondary circular reactions as "making interesting sights last" (p. 196). He speculates that infants smile and laugh at the recognition of a moderately novel event (p. 197). At the same time, it seems that they are enjoying their own power, their ability to make an event happen again and again.

Stage 4 (10 to 12 months).
The Coordination of Secondary Schemes

In stage 3, the infant performs a single action to get a result—for example, kicking to move some dangling dolls. In stage 4, the infant's actions become more differentiated; he or she learns to coordinate two separate schemes to get a result. This new accomplishment is most apparent when infants deal with obstacles. For example, one day Laurent wanted to grab a match box, but Piaget put his hand in the way. At first, Laurent tried to ignore the hand; he tried to pass over it or around it but he did not attempt to displace it. When Piaget kept his hand in the way, Laurent resorted to "storming the box while waving his hand, shaking himself, [and] wagging his head from side to side"—various "magical" gestures (1936a, p. 217). Finally, several days later, Laurent succeeded in removing the obstacle by striking the hand out of the way before he grabbed the box. Thus, Laurent coordinated two separate schemes—striking and grabbing—to obtain the goal. One scheme, striking, became a means for an end, grabbing the box.

Such simple observations are very important for our understanding of how children develop the basic categories of experience, of space and time. We cannot talk to babies and ask them about their experiences of space and time, but we can see how these categories are developing through their actions. When Laurent learned to move the hand to get the box, he showed a sense that some objects are *in front of* others in space, and that some events must *precede* others in time (Ginsburg and Opper, 1969, p. 53).

Stage 5 (12 to 18 months).
Tertiary Circular Reactions

At stage 3, infants perform a single action to obtain a single result—to make an interesting sight last. At stage 4, they perform two separate actions to obtain a single result. Now, at stage 5, they experiment with different actions to observe the different outcomes.

For example, one day Laurent became interested in a new table. He hit it with his fist several times, sometimes harder, sometimes more gently, in order to hear the different sounds that his actions produced (Piaget, 1936a, p. 270).

Similarly, a 12-month-old boy was sitting in the bathtub, watching the water pour down from the faucet. He put his hand under the faucet and noticed how the water sprayed outward. He repeated this action twice, making the interesting sight last (stage 3). But he then shifted the position of his hand, sometimes nearer, sometimes farther away from the faucet, observing how the water sprayed out at different angles. He varied his actions to see what new, different results would follow.

It is worth pausing to note that the infants were learning entirely on their own, without any adult teaching. They were developing their schemes solely out of an intrinsic curiosity about the world.

Stage 6 (18 months to two years).
The Beginnings of Thought

At stage 5, children are little scientists, varying their actions and observing the results. However, their discoveries all occur through direct physical actions. At stage 6, children seem to think out situations more internally, before they act.

The most widely known example of stage 6 behavior involves Lucienne and a matchbox. Piaget placed a chain in the box, which Lucienne immediately tried to recover. She possessed two schemes for getting the chain: turning the box over and sticking her finger in the box's slit. However, neither scheme worked. She then did something very curious. She stopped her actions and looked at the slit with great attention. Then, several times in succession, she opened and shut her mouth, wider and wider (1936a, p. 338). After this, she promptly opened the box and obtained the chain.

Piaget (1936a, p. 344) says that at stage 5 the child probably would have obtained the chain through a slow, trial-and-error process of experimenting with different actions. Because Lucienne stopped acting and thought out the situation, she was able to achieve the result much more quickly. She did not yet have a good grasp of language, so she used motor movements (her mouth) to symbolize the action she needed to perform.

Children's progress at stage 6 also can be seen in their efforts at imitation. Piaget observes that for some time children cannot imitate new models at all; they can only reproduce actions that already exist in their behavioral repertoires. By stage 5, though, they can make the necessary accommodations to imitate new behavior through experimental trial-and-error. But it is only at stage 6 that children are capable of *deferred imitation*—the imitation of absent models. For example, at 16 months of age Jacqueline

> had a visit from a little boy . . . whom she used to see from time to time, and who, in the course of the afternoon, got into a terrible temper. He screamed as he tried to get out of a play-pen and pushed it backwards, stamping his feet. J. stood watching him in amazement, never having witnessed such a scene before. The next day, she herself screamed in her play-pen and tried to move it, stamping her foot lightly several times in succession. The imitation of the whole scene was most striking (Piaget, 1946, p. 63).

Piaget says that because Jacqueline's imitation came an entire day later, she must have carried within her some internal representation of the model. Since she lacked the vocabulary to represent his actions in words, she probably used some form of motoric representation. She may have imitated his behavior with very brief muscle movements when she saw it, and these movements served as the basis for her later imitation (Piaget, 1946, Ch. 3).

The Development of Object Permanence

We have so far only described some of the main features of the six sensori-motor stages. Piaget has studied other developments during this period; he has shown how infants construct concepts of permanent objects, time, space, and causality, and how they develop capacities for play. Because of space limitations, we will briefly review only one important development—that of object permanence.

During stages 1 and 2, babies have no conception of objects existing outside themselves. If a person or an object leaves their field of vision, the most they do is to continue to look for a moment to where they last saw it. If the object does not reappear, they go on to something else. They make no attempt to search for it. For the baby, out of sight is out of mind (Piaget, 1936b, pp. 1-12).

At stage 3, new progress is made. As we mentioned earlier, babies are now becoming interested in the external world (e.g., in making interesting sights last). Consequently, they have been gaining a better sense of the permanence of external things. If objects are dropped from their line of vision, they now look to the place where the object has fallen. They also can find partly hidden objects. Also, if they temporarily put an object aside (for example, behind their back), they can, after a brief interruption, recover it. They can do so when the object was related to their own actions. However, infants at this stage cannot find objects that are completely hidden by others (1936b, pp. 13-48).

Stage 4 marks the beginning of a genuine sense of object permanence. Babies can now find completely hidden objects. If we completely cover a toy with a blanket, the baby will lift the blanket and find it (1936b, p. 51).

However, Piaget found an interesting limitation at this stage. When he hid an object at point A, his children could find it, but when he then hid the same object at point B, they again tried to find it at point A—the place of their prior success. In Piaget's terms, they could not follow a series of displacements (movements from hiding place to hiding place) (1936b, p. 54).

At stage 5, children can follow a series of displacements, so long as they can see us making them. It is only at stage 6 that infants can follow invisible displacements. For example, it was only at the sixth stage that Jacqueline could recover a ball that rolled under the sofa by making a detour around the sofa. She could do this because she now had the ability to visualize to herself, internally, the ball's trajectory path even when it was invisible (1936b, p. 231).

For Piaget, such detour behavior is very important. It shows that the child has constructed a sense of space which has the characteristics of a mathematical model called a *group*. For example, Jacqueline's detours demonstrate the principle of *associativity*, that one can reach a point through different interconnected paths. She also demonstrates the group principle of *reversibility* by bringing the ball back. Similarly, detour behavior reveals the other principles that define a coherent group structure (Piaget and Inhelder, 1966, pp. 15-17).

Less technically, we can note the tremendous progress that infants make when they achieve object permanence. At the beginning of life, they have no sense of objects existing in their own right, independent of their sight or actions. By the end of the sensori-motor period, objects are separate and permanent. Thus, children develop a universe containing independent objects, in which they are only one object among many. Along with object permanence, then, they develop a clear sense of themselves as independent beings (Piaget, 1936b, pp. 108-9).

PERIODS II AND III. PREOPERATIONAL THOUGHT (TWO TO SEVEN) AND CONCRETE OPERATIONS (SEVEN TO 11)

By the end of the sensori-motor period, the child has developed efficient and well-organized actions for dealing with the immediate environment. The child continues to use sensori-motor skills throughout life, but the next period, that of preoperational thought, is marked by a major change. The child's mind rapidly advances to a new plane, that of symbols (including images and words). As a result, the child must organize his or her thinking all over again. This cannot be done at once. For some time, during the entire preoperational period, the child's thinking is basically unsystematic and illogical. It is not until the age of seven or so, the beginning of concrete operations, that thinking becomes organized on a symbolic plane (Piaget, 1964a, p. 22).

The Growth of Symbolic Activity

Children begin to use symbols when they use one object or action to represent an absent one (Ginsburg and Opper, 1969, p. 73). Actually, as we have seen, children begin to do this during the sixth stage of sensori-motor development. For example, when Lucienne opened her mouth before opening the matchbox, she used her mouth to represent an action she had not yet performed. Similarly, deferred imitation involves some kind of internal representation—of past events. Piaget believes that deferred imitation also initially involves motoric images, and he emphasizes that the first symbols are motoric, not linguistic.

We also find examples of non-linguistic symbols in children's play. For example, one day Jacqueline pretended that a piece of cloth was her pillow. She put her head on the cloth and, laughing, pretended to go to sleep. Her play was symbolic because she used one object, a piece of cloth, to represent an absent one, the pillow (Piaget, 1946, p. 96). Make-believe play also begins during the sixth sensori-motor stage and becomes pronounced during the next few years.

A major source of symbols, of course, is language, which develops rapidly during the early preoperational years (from about two to four). One of Jacqueline's first symbolic uses of language came when she was almost two years old, after she had visited a pond. When she returned home, she told her father about the experience, saying, "Robert cry, duck swim in lake, gone away" (1946, p. 222). She thus used words to reconstruct an absent event, one from the past.

Language vastly widens the child's horizons. Through language, the child can relive the past, anticipate the future, and communicate events to others. But precisely because the young child's mind is so rapidly expanding, it initially lacks the properties of a coherent logic. This is apparent in the young child's use of words. He or she does not use words to stand for true classes of objects, but merely as *preconcepts*. For example, when Jacqueline was three years old, she said that a daddy is a man who "has lots of Luciennes and lots of Jacquelines" (1946, p. 255). She did not yet possess the concept of a general class, children, within which those with the names Lucienne and Jacqueline comprise only a small subset.

Because children lack general classes, their reasoning is frequently *transductive,* shifting from the particular to the particular. For example, at four and a half years Lucienne said, "I haven't had my nap yet so it isn't afternoon" (1946, p. 232). She did not yet understand that afternoons are general time periods which contain many particular events, of which her nap was only one.

Some psychologists believe that children learn to think more logically as they master language. In this view, language provides us with our conceptual categories (see Brown, 1965). Piaget, however, disagrees. Although language is tremendously important—it provides us with a source of shared symbols for communicating with others—it does not itself provide the structure of logical thinking. Logic, instead, stems from actions. Infants develop logically coherent action systems during the sensori-motor period, before they talk, and later logic is simply organized actions of a more internal kind (Piaget and Inhelder, 1966, pp. 86-90). To study how internal actions form logical systems, Piaget gave children various scientific tasks. He usually began such experiments with children at age four, because they could now sit down, focus on the tasks, and communicate with the examiner.

Scientific Reasoning

Conservation of continuous quantities (liquids). This is Piaget's most famous experiment. In one version (Piaget and Szeminska, 1941, p. 17), the child is shown two glasses, A1 and A2, which are filled to the same height (see Figure

5.1). The child is asked if the two glasses contain the same amount of liquid, and the child almost always agrees that they do. Next, the experimenter (or the child) pours the liquid from A2 to glass P, which is lower and wider. The child is then asked if the amount of liquid is still the same. At the *preoperational* level, the responses fall into two substages.

At the first substage, the children clearly fail to conserve—that is, they fail to realize that the quantity is the same. Usually, they say that A1 now has more because it is taller. Occasionally, the child says that P now has more because it is wider. In either case, the child "centers" on only one dimension, the height or the width. The child is so struck by a single perceptual dimension—the way it looks—that he or she fails to understand that logically the liquid must remain the same.

At the second substage, the child takes steps toward conservation but does not achieve it. A boy might at one moment say that A1 has more because it is taller, then change his mind and say that P has more because it is wider, and then become confused. The boy is showing "intuitive regulations"; he is beginning to consider *two* perceptual dimensions, but he does not yet reason about the two dimensions simultaneously and recognize that a change in one dimension cancels out a change in the other. His confusion, however, means that he is becoming aware that he is contradicting himself, and it is a good bet that he will soon resolve the contradiction and move on to the stage of conservation.

Children generally achieve conservation of liquids at about age seven. When they do so, they are entering the stage of *concrete operations.* Basically, children achieve conservation by using three arguments. First, the child might say, "You haven't added any or taken any away, so it has to be the same." This is the *identity* argument. Second, the child might say, "This glass is taller here, but the other one is wider here, so they're still the same." This is the argument of *compensation*—that the changes cancel each other out. The child assumes that the changes are part of an organized system—that a change in one dimension is necessarily related to a compensating change in another dimension. Third, the child might say, "They are still the same because you can pour this one back to what it was before." This is the argument of *inversion* (Piaget and Inhelder, 1966, p. 98). Piaget believes that the concrete operational child can use all three arguments, although the child might not spontaneously do so on any given task.

Underlying these arguments are logical *operations, mental actions* which are *reversible* (Piaget and Inhelder, 1966, p. 96). When the child argues that a change in one glass is cancelled out by a change in the other, the child understands that the end result is a return, or reversal, to the original amount. Similarly, when the child argues that we can pour the water back, he or she is suggesting that we reverse the process.

Operations, it is important to note, are mental actions. The child is carrying out compensations or reversals in his or her mind. The child has not actually performed or seen the transformations that he or she is talking about. The trans-

FIGURE 5.1 Conservation of liquid experiment. Child sees that beakers A1 and A2 contain the same amount of liquid. He then pours A2 into P, and claims that now A1 has more because it is taller.

formations (e.g., reversibility) are similar to those of the infant but are now on a more internalized plane.

People sometimes wonder if young children might fail to conserve simply because of their difficulties with language. They might think that what the experimenter means by "more" is "taller," and therefore they point to the taller glass. One can get around such difficulties by changing one's wording—for example, by asking, "Which one would give you more to drink?" Usually we find that the young child still fails to conserve (Peill, 1975, p. 7, Ch. 2).

How does the child learn conservation? The most ready answer is that conservation is taught. However, as we shall see, teaching conservation has met with unexpected resistance. The preoperational child does not genuinely believe the adult's explanations.

Piaget argues that children master conservation *spontaneously*. The crucial moment comes at the second substage, when the child first says that one glass has more because it is taller, then says the other has more because it is wider, and then becomes confused. The child is in a state of *internal contradiction,* which he or she resolves by moving on to a higher stage. Sometimes we can see this change happen before our very eyes. The child says, "This has more . . . no, that one is wider, no, wait. They're both the same. This looks taller, but you've only poured it into a wider glass."

Conservation of number. In one of his experiments on the conservation of number (Piaget and Szeminska, 1941, pp. 49-56), Piaget gave children a row of egg cups and a bunch of eggs. He then asked them to take just enough eggs to fill the cups. Again, the responses at the preoperational period fell into two substages.

At the first substage, the children simply made the rows equal in length, ignoring the number of eggs in the row. When Piaget then asked them to actually put the eggs in the cups, they were surprised to find that they had too many or too few eggs.

At the second preoperational stage, the children spontaneously created a one-to-one correspondence, placing one egg beneath each cup (see Figure 5.2). According to Piaget, they used an intuitive approach to achieve a precise perceptual order. However, their success was limited to this simple perceptual arrangement. When Piaget then bunched up (or, sometimes, spread out) one of the rows, the children claimed that now one had more. As with conservation of liquids, the children failed to conserve because they were more influenced by their immediate perceptions than by logic. Because one row now looked so much longer, they failed to reason that the number must remain the same.

At this stage, in addition, children sometimes begin to waver in their answers. One moment they say that one row has more because it is longer, but the next moment they think that the other row has more because it is denser. This state of conflict marks the transition to concrete operations.

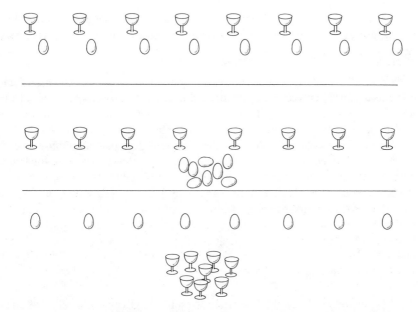

FIGURE 5.2 Conservation of number experiment. Young children often can create two rows of equal number, but if we lengthen or shorten one row, they think that the number has changed. *Adapted from* Inhelder, B. The criteria of the stages of mental development. In J. M. Tanner and B. Inhelder (Eds.), *Discussions on Child Development.* New York: International Universities Press, Inc., p. 79. Copyright © 1971 by Tavistock Publications Ltd. By permission of the publishers.

At the stage of concrete operations, children realize that the number in each row is the same despite the different appearances in length. They reason that the two rows are the same because "you haven't taken any away or added any" (identity), because "one row is longer here but this one is more bunched in" (compensation), or because "you could make this row long again and make them equal" (inversion).

Other conservative experiments. Piaget has studied several other kinds of conservation, such as the conservation of substance, weight, volume, and length. For example, in a conservation of substance experiment, the child is shown two equal balls of plasticine or play dough and then watches as one ball is rolled into a longer, thinner shape, like that of a hot dog. At the preoperational level, the child thinks that the two balls now have different amounts of play dough.

We will not describe the various kinds of conservation here, but simply note that they all are thought to involve the mastery of the same logical concepts—identity, inversion, and compensation. Nevertheless, some kinds of conservation appear more difficult than others and are mastered later.[2] Thus, the attainment of conservation is a gradual process within the concrete operational period.

Classification. In a typical classification experiment, Piaget (Piaget and Szeminska, 1941, pp. 161-81) presented children with 20 wooden beads, 18 of which were brown and two of which were white. Piaget made sure that the children understood that although most beads were brown and two were white, they all were made of wood. He then asked the children, "Are there more brown beads or more wooden beads?" At the preoperational level, the children said that there were more brown beads. Apparently they were so struck by the many brown beads in comparison to the two white ones that they failed to realize that both brown and white beads are parts of a larger whole—the class of wooden beads. As with conservation, children master such classification tasks during the period of concrete operations, and the same logical operations appear to be involved (1941, p. 178).

Social Thinking

Egocentrism. Piaget believes that at each period there is a general correspondence between scientific and social thinking. For example, just as preoperational children fail to consider two dimensions on conservation tasks, they also fail to consider more than one perspective in their interactions with others. Preoperational children are frequently egocentric, considering everything from their own single viewpoint. This is apparent from young children's conversations (Piaget, 1923). For example, a little girl might tell her friend, "I'm putting this here," oblivious to the fact that the place to which she is pointing is blocked from her friend's vision.

One of Piaget's most widely quoted studies on egocentrism dealt with the child's perception of space. In this study (Piaget and Inhelder, 1948), the child was taken for a walk around a model of three mountains so he or she could see how the model looked from different angles. After the walk, the child was seated on one side of the model, facing a doll which looked at the model from the opposite side. The child was then asked to select from among several photographs the picture that best showed what he or she saw and the picture that showed what the doll saw. All the children could pick the picture that represented their own view, but the youngest children (from about four to six years) frequently chose the same picture to show the doll's view. Apparently, they did not understand that the doll's perspective differed from their own.

Egocentrism, then, refers to the inability to distinguish one's own perspective from that of others. Egocentrism does not, however, necessarily imply

---

[2] In fact, the mastery of one series—conservation of substance, weight, and volume—seems always to occur in the same invariant sequence (Piaget and Inhelder, 1966, p. 99; Ginsburg and Opper, 1969, pp. 163-65).

selfishness or conceit. This point can be clarified by an example. One day two boys went shopping with their aunt for a birthday present for their mother. The older boy, who was seven, picked out a piece of jewelry. The younger boy, who was three and a half, selected a model car. The younger child's behavior was not selfish or greedy; he carefully wrapped the present and gave it to his mother with an expression which clearly showed that he expected her to love it. However, his behavior was egocentric; he did not consider the fact that his mother's interests were different from his own.

As long as children are egocentric, they tend simply to play alongside one another. For example, two children in the sandbox will build their own structures. As they overcome egocentrism, they learn to coordinate their actions in joint endeavors. Each might dig a tunnel so that the tunnels eventually connect. This requires considering each other's perspective. Such cooperative play occurs at the stage of concrete operations.

Piaget (1923, p. 101; 1932, p. 94) has speculated that children overcome egocentrism as they interact less exclusively with adults and more with peers. They discover that whereas grown-ups seem to understand whatever is on their minds, their peers do not. Consequently, they learn to consider others' viewpoints in order to make themselves understood.

Furthermore, children are less impressed by the authority of other children and feel freer to engage in conflicts with them. They argue with their peers and sometimes reach compromises and cooperate with them. Thus, they begin to coordinate alternative viewpoints and interests (Ginsburg and Opper, 1969, p. 94).

Whether children overcome egocentrism primarily through peer interaction or not, the most crucial point for Piaget's theory is that children themselves play an active role in grasping the fact of alternative viewpoints. On this point, I recall an instance in which one of our sons, then five years old, seemed actually to make this discovery. He was riding alone in the car with me when, after a few minutes of silence, he said, "You know, Dad, you're not remembering what I'm remembering." I asked him what he meant, and he replied, "Like I was remembering about my shoes, but you can't see what I'm remembering; you can't be remembering what I'm remembering." Thus, at that moment he seemed actually to discover, by himself, that others' perspectives differ from his own. He might not have completely surmounted his egocentrism at that instant, but the point is that whatever step he took, he took on his own.

Moral judgment. Piaget's investigations of moral behavior and thinking are contained in his classic work, *The Moral Judgment of the Child* (1932). In this study, he gave considerable attention to children's ideas about the game of marbles. He was particularly interested in whether children thought the rules could be changed.

During an early stage, lasting until about 10 years, children maintained that the rules were fixed and unchangeable. Often they said that the rules came from some prestigious authority, from the government or from God. The rules could not be changed, they asserted, because then it wouldn't be the real game.

After the age of 10 or so, the children were more relativistic. Rules were seen simply as mutually agreed-upon ways of playing the game. Children no longer considered the rules as fixed or absolute; they said the rules probably had changed over the years, as children invented new rules. And they said they too could change them, as long as everyone in the game agreed (1932, pp. 50-76).

Piaget says that these different conceptions of rules reveal two basic moral attitudes. The first, characteristic of the younger children, is moral *heteronomy*, a blind obedience to rules imposed by adults. Children assume that there is one powerful law which they must always follow. The second morality, that of the older children, is *autonomy*. This morality considers rules as human devices produced by equals for the sake of cooperation. Children achieve moral autonomy, Piaget believes, in the same way they overcome egocentrism—through interaction with peers. As they interact with peers on an equal footing, they learn that rules are not absolutes handed down from above but simply agreements which serve the purpose of cooperative interaction (1932, p. 95, pp. 401-6).

Animism. Piaget's work has suggested other ways in which young children's thinking differs from that of older children and adults. One basic difference is that children in the preoperational period do not make the same distinctions between living and non-living things that we do. Instead, their thought is often animistic; they attribute life and feelings to inanimate objects. At first, between four and six years or so, children regard anything that is active as alive. For example, when one child was asked if the sun is alive, he said yes, because "it gives light" (Piaget, 1926, p. 196). He said that a mountain is not alive because "it doesn't do anything" (1926, p. 196). A little later, between about six and eight years, children restrict life to things that move. Thus, tables and flowers are not alive because they do not move, but bicycles, rocks, and clouds are alive because they do sometimes move. Only after eight years or so do children restrict life to objects that move on their own and, later, to plants and animals.

Another aspect of animism is the attribution of feelings and consciousness to inanimate objects. For example, one seven-year-old boy was asked,

> "If you pricked a stone, would it feel it?"—*No.*—Why not?—*Because it is hard.*—If you put it in the fire, would it feel it?—*Yes.*—Why?—*Because it would get burnt...*" (1926, p. 176).

Although Piaget makes no point of it, young children's drawings also reveal animistic tendencies. For example, in the drawings in Figure 5.3, the sun and the balloon have happy faces.

Dreams. One of Piaget's earliest studies examined children's conceptions of dreams (1926, Ch. 3). As with conceptions of life, young children's under-

FIGURE 5.3   Animism in a five- and six-year-old's drawings: sun and balloon have smiling faces.

standing of dreams seems to follow a specific stage sequence. Since Piaget's first study, others (especially Kohlberg, 1966a) have refined Piaget's dream sequence.

At first children seem to believe that dreams are real. For example, when a four-year-old girl was asked if the giant in her dream was really there, she answered, "It was really there but it left when I woke up. I saw its footprint on the floor" (Kohlberg, 1966a, p. 6). Soon afterward, children discover that dreams are not real, but they still view dreams quite differently from the way older children or adults view them. They think that their dreams are visible to others and that dreams come from the outside (from the night or the sky, or through the window from the lights outside). They also think that dreams remain outside themselves while they are dreaming. It is as if they were watching a movie, with the action taking place in their rooms in front of their eyes. Gradually, stage by stage, children realize that dreams not only are unreal but are invisible, of internal origin, of internal location, and possess the other characteristics that adults assign to them. They usually complete their discoveries by the age of six or seven years, at the beginning of concrete operations.

How do children learn about dreams? Our first assumption probably is that they learn about them from adults. When children have nightmares, parents reassure them, saying, "Don't worry, it was only a dream. It wasn't real; it was only in your mind." Piagetians, however, maintain that children actually discover the various properties of dreams on their own. Kohlberg (1966a), for example, argues that because children master the dream sequence in an invariant six-stage sequence, it is unlikely that their thinking is the product of adult teachings; adults do not take the trouble to teach children about dreams in such a detailed, precise order. Instead, children arrive at different conceptualizations on their own, in an order of increasing difficulty.

To gather additional information on the role of adult teaching, Kohlberg (1966a) administered the dream interview to children in an aboriginal society in which the adults believe that dreams are real (the Atayal on Formosa). Despite the adults' beliefs, these children seem to progress through the stages in the same order as American or Swiss children. That is, they first discover that dreams are unreal, then that they are invisible, and so on. Finally, when they reach the last stage, they feel the impact of the adult views and change their minds, adopting the view that dreams are real after all. Still, they initially progress through the dream sequence in opposition to any adult beliefs, so adult views cannot be the sole determinants of their learnings.

Summary and Conclusion

Piaget argues that children's thinking during the preoperational period is very different from that of older children and adults. Preoperational thinking is characterized by egocentrism, animism, moral heteronomy, a view of dreams as external events, a lack of classification, a lack of conservation, as well as other attributes that we have not had the space to cover.

The list is a long one, and one might ask, "What do all these characteristics have in common?" The question is central to Piaget's theory, for he maintains that each developmental stage has a basic unity. Unfortunately, Piaget has not given as much attention to this question as we would like, but most often (e.g., 1964a, pp. 41-60) he tries to link the various preoperational characteristics to the concept of egocentrism.

In speech, children are egocentric when they consider matters only from their own perspective. Animism—the attribution of life to physical objects—also stems from egocentrism; children assume that everything functions just as they do. Similarly, Piaget tries to show that young children's conceptions of dreams are related to egocentrism. As long as children are egocentric, they fail to realize the extent to which each person has private, subjective experiences such as dreams. In the realm of morals, furthermore, egocentrism goes hand-in-hand with moral heteronomy. Young children regard rules from only one perspective, as absolutes handed down from above. They do not yet see how rules are based on the mutual agreements of two or more actors attempting to coordinate their different objectives in a cooperative way.

There also is a link between egocentrism and children's performances on scientific tasks, such as the experiments on conservation. Just as the egocentric child views things from a single perspective, the child who fails to conserve focuses on only one aspect of the problem. For example, when water is poured from one glass into a shorter, broader one, the child "centers" on a single striking dimension—the difference in height. The child cannot "decenter" and consider two aspects of the situation at once.

Children at the level of concrete operations are able to consider two aspects of a problem simultaneously. In their social interactions, they consider not only what they are saying, but the needs of the listener. When they perform conservation experiments, they consider not only the most visible change but compensating changes. Thus, the ability to simultaneously coordinate two perspectives forms the basis of both social and scientific thinking (Piaget, 1947, pp. 156-66).

PERIOD IV. FORMAL OPERATIONS (11 TO ADULTHOOD)

At concrete operations, children can think systematically in terms of "mental actions." For example, when water is poured into a new glass, they can tell us about the implications of reversing the process, without actually performing the activity. However, there is a limit to such abilities. They can think logically and systematically only as long as they refer to tangible objects which can be subjected to real activity (Piaget, 1964a, p. 62).

During formal operations, in contrast, thinking soars into the realm of the purely abstract and hypothetical. The capacity for abstract reasoning can be seen in responses to questions such as the following: "If Joe is shorter than Bob, and Joe is taller than Alex, who is the tallest?" At the level of concrete operations, children can handle this problem only if they actually place people in order and compare their heights; beyond this, they simply guess. At the level of formal operations, however, adolescents can order their thoughts in their minds alone (1964a, p. 62).

Piaget has been most concerned with the capacity to reason with respect to hypothetical possibilities. In one experiment (Inhelder and Piaget, 1955, pp. 107-22), children were given four flasks containing colorless liquids, labeled 1, 2, 3, and 4. They also were given a small container of colorless liquid, labeled g. Their task was to mix these liquids to make the color yellow.

At the level of preoperational intelligence, children typically made a mess. They poured the liquids in and out of the bottles in a haphazard way.

At the level of concrete operations, children's actions showed more organization. A typical strategy was to pour g into each flask: g into 1, g into 2, g into 3, and g into 4. However, they then gave up. When questioned, these children usually said that there wasn't anything more they could do. Thus, their actions revealed some organization, as we could have expected from their systematic behavior on conservation tasks, on which they can think in terms of two dimensions at once. But they entertained only a limited range of possibilities.

At the level of formal operations, the adolescents worked systematically in terms of *all possibilities*. Some started out by trying various combinations and then realized that they had better make sure that they would include all possible combinations, so they wrote them down before acting any further.

When adolescents think about the various possibilities inherent in a situation beforehand and then systematically test them, they are working like true scientists. For example, a teenage girl might decide to test the effects of a new soil for plants. At the level of formal operations, she does not just put new soil into one plant and old soil into the other and watch them grow; she considers other possibilities. Perhaps these two plants would have grown to different heights anyway, because of individual differences, so she obtains several plants and examines the average effects of the different soils. Perhaps the sunlight also has an effect—so she makes sure that all plants have the same lighting. Perhaps the amount of water also is important—so she controls for this variable too. The essence of such reasoning is that one is systematically thinking about hypotheses. One is not just entertaining a new possibility but is isolating one hypothesis by controlling for the effects of other possible variables.

As with the other periods, Piaget has introduced logico-mathematical models to describe formal operational thinking. These models are in some respects similar to those that apply to earlier developmental levels, but they also go beyond them. The models are very complex, and we will not attempt to cover them here. It is important to note, however, that at the level of formal operations, thinking reaches its highest degree of equilibrium. This means, among other things, that the various operations are more tightly interrelated and that they apply to the widest possible field of application—the realm of hypothetical possibilities.

Although Piaget has limited most of his research on adolescence to mathematical and scientific reasoning, he has speculated on the role of formal operations in the adolescent's social life (Inhelder and Piaget, 1955, Ch. 18). Unlike the concrete-operational child, who lives primarily in the here and now, adolescents begin to think about more far-reaching problems—about their futures and the nature of the society they will enter. In this process, their new cognitive powers can lead to a striking idealism and utopianism. They can now grasp abstract principles and ideals, such as liberty, justice, and love, and they envision hypothetical societies very different from any which presently exist. Thus, the adolescent becomes a dreamer, constructing theories about a better world.

Piaget thinks that such idealistic and utopian thinking carries with it a new kind of egocentrism. To fully appreciate this new egocentrism, it is necessary to review how egocentrism appears whenever the child enters a new realm of intellectual life. At first, infants are egocentric in the sense that they have no conception of the world apart from their own actions. External objects have no permanent existence of their own. Only at the end of the sensori-motor period do children decenter and situate themselves in a world of permanent objects, of which they are only one.

At the next level—that of preoperational thought—children enter a new, vastly enlarged world—one that includes language, symbolic representation, and communication with others. Children once again become egocentric and have difficulty considering more than their own immediate perspective. Gradually, they decenter and learn to consider alternative perspectives—as long as they are thinking about concrete objects immediately before them.

Finally, adolescents enter a broader world yet—the world of possibilities—and egocentrism reappears. This time egocentrism is seen when adolescents attribute unlimited power to their own thoughts. They dream of "a glorious future or of transforming the world through Ideas" (1955, p. 346), without attempting to test out their thoughts in reality. The final decentering comes about, Piaget thinks, when adolescents actually take up adult roles. They then learn the limits and resistances to their own thoughts. They learn that a theoretical construction or a utopian vision has value only in relation to how it works out in reality.

THEORETICAL ISSUES

The Stage Concept

Many psychologists use the term "stage" loosely, as merely a convenient device for summarizing their findings. This, however, is not the case with Piaget. As Kohlberg (1968) stresses, the Piagetian stage concept implies several strong positions on the nature of growth.

First, in a rigorous stage theory, the stage sequence should be invariant. People will proceed through the stages at different rates, and some may not reach the highest of Piaget's stages; but to the extent that they move through them, they will proceed in order.

Second, stages imply that growth is divided into qualitatively different periods. If intellectual development were a continuous, quantitative process, any division into separate stages would be arbitrary (Flavell, 1963, p. 19). For example, if knowledge can be scored from 0 to 100, then any division into stages at 40, 50, and 70 makes no more sense than any other series of cut-off points. Piaget, however, believes that thinking at different times is organized along qualitatively different lines. Thinking at concrete operations, for instance, is qualitatively different from that of formal operations (it is logical insofar as it refers to concrete objects and activities, but it is not yet truly abstract and hypothetical). Consequently, there is a natural, valid distinction between the two periods.

Third, stages refer to general characteristics. Kohlberg likes to discuss this point by asking the following question: "At the age of four, a child cannot copy a diamond. At the age of five, the child can. Has the child reached the diamond-copying stage?" Kohlberg explains that this proposal sounds somewhat silly because diamond-copying is too specific to be called a stage. If we were to call each particular achievement a stage, we would have thousands of stages. It is more appropriate to say that the child has reached a new *general* stage of perceptual-motor coordination that permits him or her to do many new things.

Similarly, Piaget's stages refer to general patterns of thought, and if we know that a child is in a particular stage, we should be able to predict the child's behavior across a wide variety of tasks. This is not completely true, for children may be at somewhat different stages in different areas (e.g., in scientific versus social reasoning). However, there should be substantial unity in performances at each general period.

Fourth, Piaget, like other rigorous stage theorists, claims that his stages unfold in the same sequence in all cultures. This proposal frequently puzzles readers. Don't different cultures teach different beliefs, particularly with regard to morals? We will take up this issue in the next chapter, but in general the Piagetian answer is that the theory is not concerned with specific beliefs but with underlying cognitive abilities. Thus, young children, regardless of their cultural beliefs on such matters as sex or fighting, will base their views on what they think authority condones or punishes. It is not until adolescence, when young people acquire formal operations, that they will give abstract, theoretical treatises on moral matters, whatever their specific views.

In summary, then, Piaget advances a rigorous stage theory, which means that he believes his stages a) unfold in an invariant sequence, b) describe qualitatively different patterns, c) refer to general properties of thought, and d) are culturally universal.[3]

Movement from Stage to Stage

Piaget has devoted a great deal of attention to the structures of his stages and far less attention to the problem of movement through them. Nevertheless, he has definite views on this topic.

He (1964b) acknowledges that biological maturation plays some role in development. For example, children probably cannot attain concrete operations without some minimal maturation of the nervous system. However, Piaget says that maturation alone cannot play the dominant role because rates of development depend so much upon where children live. Children who grow up in impoverished rural areas frequently develop at slow rates, apparently because they lack intellectual stimulation. Thus, the environment also is important.

However, it is easy to exaggerate the role of the environment, as learning theorists do. Generally speaking, learning theorists believe that the child's mind is primarily a product of external reinforcements and teaching. Piagetian concepts, they assume, must be taught by parents, teachers, and others. However, it is not all that clear that this is the case. Researchers have had only limited success in teaching conservation, for example (Liebert *et al.,* 1977, pp. 176-79).

[3] Sometimes Piagetians say that stages are hierarchically integrated, which means that lower stages do not disappear, but become integrated into higher ones (Inhelder, 1971, p. 85). However, it does not appear that most features of preoperational thought become integrated into any new stage structure; they are simply overcome.

In Piaget's view, the environment is important, but only partly so. The environment nourishes, stimulates, and challenges the child, but children themselves build cognitive structures. As children seek out the environment, they encounter events that capture their *interest*. These events are moderately novel (Piaget, 1936a, p. 68), events that do not quite fit into their existing structures. Children then adjust their actions to learn about them, and in the process build new ways for dealing with the world. For example, we saw how a little boy placed his hand under the faucet and was struck by the way the water sprayed outward. He then adjusted his hand up and down to learn more about it, and as he did so, he learned a little about the efficacy of actively experimenting with different actions to see different effects (stage 5 of sensori-motor development). In such behavior, it is not the environment that structures the child's mind but the child who develops new cognitive schemes.

Experiences that promote cognitive development, in addition, are not only interesting, but usually place the child in a state of *conflict*. For example, an infant might be unable to grasp an object because an obstacle is in the way. The child needs to invent a new structure—a means-ends relationship—to obtain the object. The child assimilates new objects by making accommodations which build new cognitive structures.

The concept of conflict is involved in a fairly recent model of development change which Piaget calls equilibration (1964b). We already have discussed the essence of this model, without using its name, when we described how children achieve conservation. For example, a little girl sees a ball of clay elongated and initially thinks that the amount has increased. After a while, however, she considers the clay's narrow width and thinks the clay has shrunk. Thus, she perceives something that contradicts her initial view. When she thinks about both the length and the width, she becomes confused. This conflict motivates the child to realize that one change cancels out the other, leading to the discovery of conservation. Piaget's equilibration model tries to assign numerical probabilities to the chances that the child will consider one dimension, then the other, and finally both.

Another source of new, conflicting information is the social environment. For example, preoperational children overcome egocentrism when they interact with peers, with whom they get into arguments and conflicts. In such interchanges, they learn that others have views different from their own, and they also learn to coordinate different interests to behave in a cooperative fashion. This ability to coordinate viewpoints may aid in the growth of scientific thinking, where the coordination of dimensions also is important (Piaget, 1947, pp. 156-66).

Piaget, then, has tried to indicate different ways in which interesting and conflicting pieces of information lead children to develop new cognitive structures. It is important to emphasize that development is always a spontaneous process. It is the children themselves who assimilate new information, who make discoveries, and who resolve inner contradictions.

IMPLICATIONS FOR EDUCATION

Piaget has not written extensively on education, but he does have some recommendations. Essentially, his overall educational philosophy is similar to that of Rousseau and Montessori. For Piaget, too, true learning is not something handed down by the teacher, but something that comes from the child. It is a process of active discovery. This is clearly true of infants, who make incredible intellectual progress simply by exploring the world on their own, and it can be true of older children as well. Accordingly, the teacher should not try to impose knowledge on the child, but he or she should find materials that will interest and challenge the child and then permit the child to solve problems on his or her own (Piaget, 1969, pp. 151-53, 160).

Like Rousseau and Montessori, Piaget also stresses the importance of gearing instruction to the child's particular level. He does not agree with Montessori's maturational view of stages, but the general principle still holds: The educator must appreciate the extent to which children's interests and modes of learning are different at different times. Say, for example, a boy is just entering the stage of concrete operations. He is beginning to think logically, but his thinking is still partly tied to concrete objects and activities. Accordingly, lessons should give him opportunities to deal actively with real things. If, for example, we wish to teach him about fractions, we should not draw diagrams, give him lectures, or engage him in verbal discussions. We should allow him to divide concrete objects into parts (Flavell, 1963, p. 368). When we assume that he will learn on a verbal plane, we are being egocentric; we are assuming that he learns just as we do. The result will be a lesson which sails over his head and seems unnatural to him.

It might appear that this principle—tailoring education to the child's own stage—is self-evident. Unfortunately, however, this is not always so. A case in point, according to Kohlberg and Gilligan (1971, p. 1082) is the wave of curricular reforms referred to as the "new math," "new science," and "new social studies." The objective, in the case of new math, is to go beyond rote learning of mechanical skills and to teach the theories and ideas behind the operations. At first glance, this appears to be a noble ambition. However, new math has not worked very well. The reason, according to Kohlberg, is that it attempts to teach young children, who are largely at the level of concrete operations or lower, ideas that assume capacities at the abstract level of formal operations. Thus, new math begins with an adult conception of what children should learn and ignores children's cognitive levels. Careful attention to Piaget's stages might help educators find more appropriate starting points for instruction.

It is not always easy to find the right educational experiences for a given child. A knowledge of cognitive stages can help, but children are sometimes at different stages in different areas (Piaget, 1969, p. 171). What is needed is sensitivity and flexibility on the teacher's part—a willingness to look closely at the child's actions, to learn from the child, and to be guided by the child's spontaneous interests (Ginsburg and Opper, 1969, p. 220). For active learning always presupposes interest (Piaget, 1969, p. 152).

Like Rousseau and Montessori, then, Piaget thinks learning should be a process of active discovery and should be geared to the child's stage. However, Piaget disagrees with Rousseau and Montessori on one point. Piaget sees much greater educational value in social interactions. Children begin to think logically— to coordinate two dimensions simultaneously—partly by learning to consider two or more perspectives in their dealings with others. Thus, interactions should be encouraged, and the most beneficial ones are those in which children feel a basic equality, as they most often do with peers. As long as children feel dominated by an authority who knows the "right" answer, they will have difficulty appreciating differences in perspectives. In group discussions with other children, in contrast, they have a better opportunity to deal with different viewpoints as stimulating challenges to their own thinking (Piaget, 1969, pp. 173-80).

EVALUATION

Research

As mentioned earlier, Piaget's work was neglected for many years and this was partly because of its scientific shortcomings (e.g., his failure to report results in statistical form). Nevertheless, Piaget forged ahead, making discovery after discovery and building an overall theory. Finally, in the 1960s, psychologists recognized the importance of Piaget's work, and it has stimulated so much sound research that any criticisms of its methodology are no longer relevant.

The research bearing on Piaget's theory is far too vast to summarize here, but we can mention some general trends, issues, and questions.

On the whole, Piaget's *stage sequence* appears solid. That is, children do seem to move through the substages, stages, and periods in the order Piaget initially found. His stages have held up particularly well for the sensori-motor period and for scientific and mathematical reasoning with respect to the later stages (Lovell, 1968; Corman and Escalona, 1969; Almy *et al.*, 1966; Flavell, 1977; Neimark, 1975; Dasen, 1972). There have been more controversies over his stages of social thought, such as animism (Jahoda, 1958), moral judgment (Kohlberg, 1964), and egocentrism (Borke, 1975; Chandler and Greenspan, 1972; Flavell *et al.*, 1968). For example, whereas young children (e.g., four-year-olds) almost invariably fail to conserve, they do not seem invariably animistic or egocentric. They only seem more egocentric or animistic than older children on the average.[4]

Some researchers have questioned Piaget's claim that his stages represent *general* modes of thinking across a wide variety of tasks. Their studies have found the correlations among tasks to be rather low, even when the tasks are in the same area of thinking (Flavell, 1977, p. 248; Sigel, 1968, pp. 507-8). Intercorrelations do seem higher, though, when one examines children who are firmly within a general period, rather than those who are in a transitional state (Uzgiris, 1964).

One of the most perplexing findings is the general failure of most adults to use formal operations in any consistent way. Most middle-class American adults do not employ the highest stages of formal reasoning on most standard tasks (Neimark, 1975; Flavell, 1977, p. 114; Kohlberg and Gilligan, 1971), and in many non-Western cultures many adults barely use any formal operations at all (Dasen, 1972). In a recent essay (1972), Piaget has attempted to account for such findings. It may be, he says, that adults generally attain some degree of formal operational thinking, if not by age 15, by age 20 or so, but they employ formal operations primarily in areas of special interest or ability. For example, an automobile mechanic may not think in a formal, theoretical way about philosophy, physics, or literature, but may use formal operations when troubleshooting a car. Similarly, Tulkin and Konner (1973) have suggested that preliterate peoples may fail to demonstrate formal operations on Piagetian tasks of mathematical and scientific reasoning but employ them when working on problems of urgent importance to them. For example, when Kalahari bushmen in Africa discuss animal tracking, they advance and weigh hypotheses in ways "that tax the best inferential and analytic capacities of the human mind" (p. 35).

Perhaps Piaget's most controversial claim is that cognitive development is a spontaneous process. Children, he says, develop cognitive structures on their own, without direct teaching from adults. To many psychologists, particularly American learning theorists, this claim must surely be wrong. Accordingly, numerous researchers have designed "training studies," most of which have tried to teach conservation to four- and five-year-olds.

A major finding is that conservation is surprising difficult to teach (Flavell, 1963, p. 377; Liebert *et al.*, 1977, pp. 176-79). It is difficult to teach conservation by simply explaining and reinforcing the right answers. And if one does succeed on one task, the ability does not always generalize to new tasks. Further, the teaching does not always cut very deep. For example, people have told me how

---

[4] A few studies (e.g., Bower, 1976) suggest that an interesting revision in Piaget's stage sequence may be necessary. Children might briefly show abilities, such as conservation, far ahead of schedule (e.g., at age two), and then lose the ability until they relearn it. Such findings, however, are extremely preliminary.

they had apparently taught a child to conserve liquids; however, when they then offered the child a choice between liquids he or she liked to drink, the child insisted on taking the larger glass.

Nevertheless, it does now seem that one can teach conservation, if one works at it hard enough. In the most successful experiment, Gelman (1969) taught children to conserve number and length by reinforcing them for attending to the most relevant stimuli—for example, the number of objects in a row rather than the row's length. Further, 60 percent of the children showed an immediate new ability to conserve substance and liquid. However, Gelman's training procedure was laborious. The training lasted two days and consisted of 192 trials. One wonders whether such methods accurately reflect the ways in which children master conservation in their daily lives. One can also wonder about the effect of such training on the children's feelings. When children solve problems on their own they gain confidence in their abilities to make discoveries. When they undergo an intensive training program, in which they are consistently reinforced for responding in a manner in which they would not ordinarily respond, they might learn to mistrust their own powers of thinking.

Theoretical Comment

Piaget is stimulating because he challenges our common assumptions about children. In a sense, his most radical ideas are as old as Rousseau, but they might easily have been lost had it not been for Piaget's work.

Fundamentally, Piaget's work has helped give substance to two ideas. First, children are not just taught by adults; they learn on their own. The most incontestable evidence for spontaneous learning comes from Piaget's observations on infants, who make enormous intellectual progress simply by exploring the environment on their own, before anyone takes the trouble to educate them. Once we begin teaching, in fact, we often seem to stifle the child's natural curiosity. In school, children become disinterested, lazy, rebellious, and frightened of failure. The major task of education, it would seem, would be to liberate the bold curiosity with which children enter life.

Piaget's second fundamental concept is that children think differently from adults. They do not merely know less; they think in an entirely different way. This seems most true of the preoperational child, who considers dreams, morals, life, and scientific matters very differently from adults. Once we accept Piaget's assumption—that children think differently from us—we begin to adopt a new attitude toward them. We become more open-minded. We do not assume that we know what is on children's minds, but we let them reveal themselves to us. We become like explorers in a new, interesting territory.

At the same time, however, Piaget's own theoretical apparatus sometimes gets in the way of a completely open appreciation of childhood. For one thing, he sometimes seems more interested in the intricacies of logic than in children. Amidst all the logical analysis, the child often gets lost (Flavell, 1963, pp. 427-28).

Moreover, I wonder whether Piaget provides a full appreciation of one of his most important stages—that of preoperational thought. This period is often described in a one-sided, negative manner. The preoperational child is always failing. The child "continues to make the same mistake," "cannot make coordinations," and "fails to grasp" basic notions (Piaget and Szeminska, 1941, p. 142, 77, 13). Preoperational thought, Piaget repeatedly says, is prelogical, irreversible, static and perception-bound, full of contradictions and errors, and so forth. While it is true that the preoperational child takes steps toward concrete operational thinking, preoperational thought, as Flavell says, "can scarcely be called 'good' thought" (1963, p. 156).

Of course, Piaget is not trying to make value judgments. He is simply attempting to give an accurate account of what the child lacks. But can a full appreciation of preoperational thinking be so consistently negative? Might not there also be valuable aspects of this thinking that Piaget overlooks? Theorists in the Romantic tradition have always been open to the possibility that each period of childhood has its own special virtues which may become lost as the child becomes an adult.

One theorist who had interesting things to say on this topic was Heinz Werner, a developmental psychologist whose own theory is in many ways similar to Piaget's. In Werner's view (1948), there are different kinds of thinking, each with its own course of growth. One kind is "geometrical-technical" intelligence, which includes logical and scientific reasoning—the kinds of thinking that Piaget has most fully explored. Geometrical-technical reasoning increasingly dominates the thinking of the growing child in advanced, industrialized cultures. Another kind of thinking and perception is called "physiognomic" and is dominant in early childhood. The child who perceives the world physiognomically does not differentiate between animate and inanimate objects and therefore tends to perceive the whole world as alive and full of feeling. Physiognomic perception is similar to what Piaget calls animism. When, for example, the child thinks that a rock feels pain when we burn it, the child is thinking physiognomically.

According to Werner, physiognomic perception not only characterizes young children's orientation but that of preliterate peoples and regressed schizophrenics, for they too think primarily in "primitive" ways. But physiognomic perception also characterizes the world of the artist and is a source of artistic creativity. For example, the painter Kandinsky writes:

> On my palette sit high, round rain-drops, puckishly flirting with each other, swaying and trembling. Unexpectedly they unite and suddenly become thin, sly threads which disappear in amongst the colors, and roguishly skip about and creep up the sleeves of my coat. . . . It is not only the stars which show me faces. The stub of a cigarette lying in an ash-tray,

a patient, staring white button lying amidst the litter of the street, a willing, pliable bit of bark—all these have physiognomies for me . . ." (Werner, 1948, p. 71).

Kandinsky, then, perceives life and feeling in inanimate things—in his paint, in cigarettes, in bark. In particular, he sees faces (physiognomies) in objects, much like the child who draws a face in the sun or on the front of a car. This excerpt reveals how foreign physiognomic perception has become to us as adults. If we did not know that Kandinsky was an artist, we surely would consider him mad.

Werner believes that physiognomic perception, while submerged in our technological cultures, is as valuable as geometrical-technical intelligence. If we lose this way of thinking about the world, we lose something valuable—the artist within ourselves. The ideal human, Werner thinks, would be someone who has developed *both* geometrical-technical and physiognomic modalities. It would be someone who has advanced to the heights of formal operational capabilities (to use Piaget's term), but someone who also has a receptivitiy to earlier, animistic and preoperational modes of experience.

Werner's speculations, then, help place Piaget in perspective. Piaget has systematically mapped out the growth of logical and scientific thinking. In terms of this kind of thinking, animism and other qualities of preoperational thought are, indeed, inferior ways of thinking. Werner, however, does not consider preoperational thinking so negatively. Preoperational thinking may be the kind that artists utilize and therefore may be worth preserving and developing. Many artists have said that they try to recover and to nurture the kinds of attitudes they had as children (Gardner, 1973, pp. 20-21). In some ways, then, the preoperational child, while an inferior logician, is a superior artist. Piaget has paid too little attention to the virtues of this side of life.

Kohlberg's Stages of Moral Development

6

BIOGRAPHICAL INTRODUCTION

One of the outstanding examples of research in the Piagetian tradition is the work of Lawrence Kohlberg. Kohlberg has focused on moral development and has proposed a stage theory of moral thinking which goes well beyond Piaget's initial formulations.

Kohlberg, who was born in 1927, grew up in Bronxville, New York, and attended the Andover Academy in Massachusetts, a private high school for bright and usually wealthy students.[1] He did not go immediately to college, but instead went to help the Israeli cause. There he was made the Second Engineer on an old freighter carrying refugees from parts of Europe to Israel. After this, in 1948, he enrolled at the University of Chicago, where he scored so high on admission tests that he had to take only a few courses to earn his bachelor's

---

[1] I would like to thank David F. Ricks for his assistance with this biographical section.

degree. This he did in one year. He stayed on at Chicago for graduate work in psychology, at first thinking he would become a clinical psychologist. However, he soon became interested in Piaget, and began interviewing children and adolescents on moral issues. The result was his doctoral dissertation (1958a), the first rendition of his new stage theory.

Kohlberg is an informal, unassuming man who also is a true scholar; he has thought long and deeply about a wide range of issues in both psychology and philosophy and has done much to help others appreciate the wisdom of many of the "old psychologists," such as Rousseau, John Dewey, and James Mark Baldwin. Kohlberg has taught at the University of Chicago (1962-1968) and, since 1968, has been at Harvard University.

PIAGET'S STAGES OF MORAL JUDGMENT

Piaget studied many aspects of moral judgment, but most of his findings fit into a two-stage theory. Children younger than 10 or 11 years think about moral dilemmas one way; older children consider them differently. As we have seen, younger children regard rules as fixed and absolute. They believe that rules are handed down by adults or by God and that one cannot change them. The older child's view is more relativistic. He or she understands that it is permissible to change rules if everyone agrees. Rules are not sacred and absolute, but are devices which humans use to get along cooperatively.

At approximately the same time—10 or 11 years—children's moral thinking undergoes other shifts. In particular, younger children base their moral judgments more on consequences, whereas older children base their judgments on intentions. When, for example, the young child hears about one boy who broke 15 cups trying to help his mother and another boy who broke only one cup trying to steal cookies, the young child thinks that the first boy did worse. The child primarily considers the amount of damage—the consequences—whereas the older child is more likely to judge wrongness in terms of the motives underlying the act (Piaget, 1932, p. 137).

There are many more details to Piaget's work on moral judgment, but he essentially found a series of changes that occur between the ages of 10 and 12, just when the child begins to enter the general stage of formal operations.

Intellectual development, however, does not stop at this point. This is just the beginning of formal operations, which continue to develop at least until age 16. Accordingly, one might expect thinking about moral issues to continue to develop throughout adolescence. Kohlberg, then, interviewed both children and adolescents about moral dilemmas.

KOHLBERG'S METHOD

Kohlberg's (1958a) core sample was comprised of 72 boys, from both middle- and lower-class families in Chicago. They were ages 10, 13, and 16. He later added to his sample younger children, delinquents, and boys and girls from other American cities and from other countries (1963, 1970).

The basic interview consists of a series of dilemmas such as the following:

Heinz Steals the Drug

In Europe, a woman was near death from a special kind of cancer. There was one drug that the doctors thought might save her. It was a form of radium that a druggist in the same town had recently discovered. The drug was expensive to make, but the druggist was charging ten times what the drug cost him to make. He paid $200 for the radium and charged $2,000 for a small dose of the drug. The sick woman's husband, Heinz, went to everyone he knew to borrow the money, but he could only get together about $1,000 which is half of what it cost. He told the druggist that his wife was dying and asked him to sell it cheaper or let him pay later. But the druggist said: "No, I discovered the drug and I'm going to make money from it." So Heinz got desperate and broke into the man's store to steal the drug for his wife. Should the husband have done that? (Kohlberg, 1963, p. 19).

Kohlberg is not really interested in whether the subject says "yes" or "no" to this dilemma but in the reasoning behind the answer. The interviewer wants to know *why* the subject thinks Heinz should or should not have stolen the drug. The interview schedule then asks new questions which help one understand the child's reasoning. For example, children are asked if Heinz had a right to steal the drug, if he was violating the druggist's rights, and what sentence the judge should give him once he was caught. Once again, the main concern is with the reasoning behind the answers. The interview then goes on to give more stories, to get a good sampling of a subject's moral thinking.

Once Kohlberg had classified the various responses into stages, he wanted to know whether his classification was *reliable*. In particular, he wanted to know if others would score the protocols in the same way. Other judges independently scored a sample of responses and he calculated the degree to which all raters agreed. This procedure is called *interrater reliability*. Whenever investigators use Kohlberg's interview, they also should check for interrater reliability before scoring the entire sample.[2]

[2] There are other forms of reliability which also are important (see Kurtines and Grief, 1974).

KOHLBERG'S SIX STAGES

Level I. Preconventional Morality

Stage 1: Obedience and Punishment Orientation. Kohlberg's stage 1 is similar to Piaget's first stage of moral thought. The child assumes that powerful authorities hand down a fixed set of rules which he or she must unquestioningly obey. To the Heinz dilemma, the child typically says that Heinz was wrong to steal the drug because "It's against the law," or "It's bad to steal," as if this were all there were to it. When asked to elaborate, the child usually responds in terms of the consequences involved, explaining that stealing is bad "because you'll get punished" (Kohlberg, 1958b).

Although the vast majority of children at stage 1 oppose Heinz's theft, it is still possible for a child to support the action and still employ stage 1 reasoning. For example, a child might say, "Heinz can steal it because he asked first and it's not like he stole something big; he won't get punished" (see Rest, 1973). Even though the child agrees with Heinz's action, the reasoning is still stage 1; the concern is with what authorities permit and punish.

Stage 2: Relativistic Hedonism. At this stage, children no longer consider rules as so fixed and absolute. They realize that there is more than one side to any issue, and they therefore become relativistic. Stage 2 children frequently say things as, "It all depends on how you look at it," and "It depends on the person. Heinz might think it is right to steal to save his wife, but the druggist might think it's wrong." Because everything is relative, one ultimately decides according to one's needs and pleasure—hedonistically. A stage 2 respondent might say that Heinz can steal if he loves or needs his wife, but he doesn't have to if, say, he wishes to marry someone else who is younger and better-looking (Kohlberg, 1963, p. 24).

Kohlberg's second stage is similar to Piaget's second stage; in both there is a new relativism. However, in some cases Kohlberg found an amoral cynicism at this stage which was not present in Piaget's interview protocols.

Level II. Conventional Morality

Stage 3: Good Boy/Good Girl Orientation. At this stage, the child—who by now is usually entering his or her teens—responds in terms of what a "good" person would do. Goodness is defined in terms of *motives* and *feelings*. The young person emphasizes that Heinz was "trying to save a life," that "he loved his wife," and that "he was desperate and that was the only reason he stole." Conversely, the druggist's motives were "bad": He was "greedy," he "was only out to make a profit," and he "was only interested in himself, not another life" (Kohlberg, 1958b). Sometimes the youngster is so angry with the druggist that he or she says that the druggist ought to be put in jail.

A typical stage 3 response is that of Don, age 13:

> It was really the druggist's fault, he was unfair, trying to overcharge and letting someone die. Heinz loved his wife and wanted to save her. I think anyone would. I don't think they would put him in jail. The judge would look at all sides, and see that the druggist was charging too much (Kohlberg, 1963, p. 25).

We see that Don defines the issue in terms of the actors' character traits and motives. He talks about the loving husband, the unfair druggist, and the understanding judge. His answer deserves the label "conventional morality" because it assumes that the attitude expressed would be shared by the entire community— "anyone" would be right to do what Heinz did in his place (Kohlberg, 1963, p. 25).

We have observed that Kohlberg's first two stages roughly parallel Piaget's. In both cases, the child at stage 1 views rules as fixed and absolute and at stage 2 sees them more relativistically. However, there also is a difference. Piaget thought that children begin reasoning in terms of intentions rather than consequences at stage 2; for Kohlberg, this shift occurs at stage 3. The comparison is diagrammed in Figure 6.1.

	Piaget	**Kohlberg**
Stage 1	Unquestioning obedience to authority	(Similar)
	Judgments in terms of consequences	(Similar)
Stage 2	Relativistic view of rules	(Similar)
	Judgments in terms of intentions	
Stage 3		Judgments in terms of intentions (meeting with community's approval)

FIGURE 6.1   Kohlberg's first three stages compared with Piaget's stages.

Stage 4: Maintenance of Social Order and Authority. The stage 3 respondent is forming an image of community approval. Right conduct consists of well-meaning behavior that every good and understanding person would approve of. At stage 4, the young person thinks more broadly about the general social order. Now the emphasis is on obeying laws so that the social order will be maintained. In response to the Heinz story, the young person is usually sympathetic to Heinz but cannot condone the theft. What would happen if we all started breaking the laws whenever we felt we had a good reason? The result would be chaos.

It is rather difficult to find pure stage 4 reasoning on the Heinz story (in comparison to other dilemmas) because most people at this stage also are impressed by the importance of saving a life. This is one point where we frequently find mixtures of stages. For example, we might find a mixture of stage 3 and stage 4 thinking. The respondent might say that Heinz should steal because he loves his wife and is in desperate straits (stage 3), but Heinz also should pay the druggist back and go to jail, for the law must be respected as the crucial element in the social fabric (stage 4).

You will recall that stage 1 children also generally oppose stealing because it breaks the law. Superficially, stage 1 and stage 4 subjects are giving the same response, so we see here why Kohlberg insists that we must probe into the reasoning behind the initial overt response. Stage 1 children say, "It's wrong to steal" and "It's against the law," but they cannot elaborate any further, except to say that stealing can get a person jailed. Stage 4 youngsters, in contrast, have a conception of the function of laws for society as a whole. This conception far exceeds the grasp of the younger child. Stage 4 respondents talk about the "need for order," the importance of "everyone fulfilling his duties and obligations," and the importance of "preventing chaos."

Level III. Postconventional Morality

Stage 5: Democratically Accepted Law. At stage 4, the individual takes a fairly rigid "law and order" stance. Adherence to the law and to the social order has a frozen quality. At stage 5, in contrast, the individual views laws more flexibly. Laws are devices which the community agrees upon so that people can live in harmony. If people feel the laws are not meeting their needs, they can always change them through mutual agreement and democratic procedures.

The stage 5 respondent, in addition, has some glimmering of personal values which may even be higher than the law. These values might include liberty, justice, and the pursuit of happiness. If anyone feels that the laws are not serving his or her personal values, this individual has the right to persuade others to change the laws. Still, the stage 5 thinker is not an advocate of civil disobedience. The individual does not advocate breaking a law because some higher ethical principle is at stake. One has a right to try to change laws, but one must do so in an orderly, democratic fashion.

In response to the Heinz story, the stage 5 respondent is usually in a quandary. On the one hand, the individual has a deep, rational commitment to the law, so stealing is wrong. On the other hand, the respondent dimly perceives that a higher principle is involved in this case—everyone's right to a fulfilling life, which includes the wife's right to live. But the respondent cannot formulate this value in a way that makes it right for Heinz to steal, so he or she seems confused. For example, one 16-year-old said, "In my eyes he'd have just cause to do it, but in the law's eyes he'd be wrong. I can't say more than that as to whether it was right or not" (Kohlberg, 1963, p. 29). Thus, this boy has some recognition of Heinz's "just cause," which transcends the law, but he cannot give a clear and comprehensive formulation of it.

Stage 6: Universal Principles. At stage 6, individuals have a clearer conception of certain abstract, universal principles which transcend the law. These principles include justice for all and the dignity of every individual. Stage 6 thinkers recognize the importance of social order, but they also realize that not all orderly societies fulfill more important principles.

In response to the Heinz story, stage 6 subjects say that Heinz had no legal right to steal to save his wife, but he had a higher, moral right. For every individual has an absolute worth—each life has an inherent value (Kohlberg, 1970). This principle is universal; it applies to all. If one holds that individuals possess an absolute worth, one cannot turn around and say this person is of value, but not that one. Thus, we have an obligation to save anyone, not just someone we love. We might not always do so, but the obligation is there.

Universal ethical principles are difficult to formulate, and few people reach stage 6 moral thinking. The discovery of universal principles has been the goal of the great moral and religious leaders and philosophers, such as Gandhi, Jesus, Socrates, Martin Luther King, and Kant. Martin Luther King once said in a television interview, for example, that he believed in universal brotherhood and love. The interviewer looked puzzled, and asked, "Do you mean you believe in loving those racists who abuse you and who are trying to stop your cause?" King's reply was something like, "Yes, I said I believe this is a universal value. If I say I believe in love for all mankind, I mean all mankind." King felt the inner need to be logically consistent and comprehensive with respect to the principle he had chosen.

Similarly, the philosopher Immanuel Kant (1788) spent a lifetime searching for a principle which would have universal application—a principle according to which every individual could lead his or her life, regardless of the circumstances. He called such a principle the "categorical imperative," and he tried out various alternatives. One version was that we should always treat others as an end, never as a means. For example, we should not pretend to be someone's friend only to advance our own social status and then stop the friendship when we get where we want. Instead, we should treat others as ends, as important in their own right, and consider their feelings.

Other candidates for universal principles include the Golden Rule (do unto others as you would have them do unto you), the sacredness of life, the principle of the greatest good for the greatest number, and the principle of fostering human growth. It is not clear that anyone has found a universal principle that works in all situations, but this is the goal. Kohlberg himself believes that all universal principles involve a concept of justice. Justice means that we reach a solution that everyone considers fair. A fair solution comes about when everyone impartially puts himself or herself in everyone else's shoes (Kohlberg and Elfenbein, 1975). In the case of the Heinz conflict, even the druggist would realize that if he were in the wife's predicament, he would value his life over any profit, so a just solution would include a recognition of this higher value. Of course, if the druggist considered himself more important than the wife, he might let her die anyway. Justice, as a universal principle, implies that everyone is considered equal.

Kohlberg says that stage 6 principles, such as justice or the basic dignity of all individuals, are self-chosen. Unlike the Ten Commandments, their authority does not come from God or any other external source. We freely choose certain ethical principles because they seem to lead to the best life for all. At the same time, if one is to be consistent and if one's choices mean anything, one has an obligation to live up to one's principles.

Summary

At stage 1, children assume that one must unquestioningly obey authority, or one will get punished. At stage 2, children are no longer so impressed by any single authority; they see that there are different sides to every issue. As long as everything is relative to the person's viewpoint, one is free to make decisions according to one's wishes—hedonistically.

At stages 3 and 4, young people judge conduct in terms of the conventional society. At stage 3, children assume that the good members of the community will approve of one's behavior so long as one's motives are pure. At stage 4, the main concern is with adherence to the laws for the sake of the social order.

At stages 5 and 6, the person considers rights and principles that may override the society's values and its need for order. At stage 5, the respondent has some inkling of personal values that may take priority over the law, whereas at stage 6 the individual conceptualizes such values as abstract, universal principles.

THEORETICAL ISSUES

Invariant Sequence

Kohlberg, like Piaget, believes that his stages unfold in an invariant sequence. Children always go from stage 1 to stage 2 to stage 3 and so forth. They do not skip stages or move through them in mixed-up orders. Not all children eventually attain the highest stages, but to the extent that they do go through them, they proceed in order.

Most of Kohlberg's evidence on his stage sequence comes from *cross-sectional* data. That is, he interviewed different children at various ages to see if the younger ones were at lower stages than the older children. Figure 6.2 summarizes this data from his first studies. As you can see, stages 1 and 2 are primarily found at the youngest age, whereas the higher stages become more prevalent as age increases. Thus, the data support the stage sequence.

Cross-sectional findings, however, are inconclusive. In a cross-sectional study, different children are interviewed at each age, so there is no guarantee that any individual child actually moves through the stages in order. For example, there is no guarantee that a boy who is coded at stage 3 at age 13 actually passed through stages 1 and 2 in order when he was younger. More conclusive evidence must come from *longitudinal* studies, in which the same children are followed over time.

We do have a few longitudinal studies, but some of the results are ambiguous.

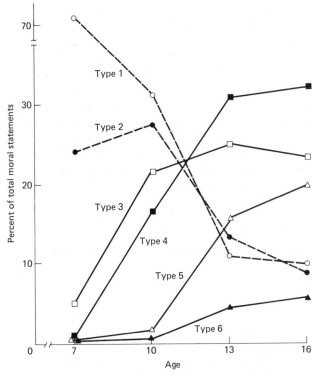

FIGURE 6.2   Use of six types of moral judgments at four ages.

Source: Kohlberg, L. (1963). The development of children's orientations toward a moral order: I. Sequence in the development or moral thought. Human Development, 6, p. 16. By permission of the publisher.

Two studies (Kohlberg and Kramer, 1969; Holstein, 1973) began with samples of teenagers and then tested them at three-year intervals. In both, most subjects either remained at the same stage or moved up one stage, but there were some who might have skipped over a stage. Evidence of stage skipping would damage Kohlberg's theory, but we cannot be sure that it really occurred. Because the interval was three years, it is quite possible that some subjects moved two stages forward during this period. There is a third study (Kuhn, 1976) which tested younger children at six-month intervals and found more orderly progressions, with no evidence of stage skipping. However, we also need data on older children and adolescents as well.

The logical order of Kohlberg's sequence. Although there is no substitute for empirical data, there does seem to be a logical sense in which Kohlberg's stages should follow one another because they represent increasingly differentiated and abstract ways of judging moral matters. At stage 1, children view moral issues from a single perspective, considering only external authority. At stage 2, children realize that there are different views on any issue; their thinking becomes more differentiated. The transition from stage 1 to 2 seems to involve the overcoming of egocentrism. The child at stage 2 is aware of different viewpoints on moral problems.

Stage 3 represents a further cognitive advance. Children grasp qualities that are more abstract and intangible—motives and intentions underlying behavior. They also begin to consider wider community values, assuming, for example, that every good person would approve of an act if the motives are pure. Kohlberg thinks that stage 3 thinking requires some formal operations (Kohlberg and Gilligan, 1971). Stage 4 children have a grasp of a broader concept yet—society as a whole.

At stage 5, young people further see how laws are based on community agreements and have some inkling of values which may go beyond society's laws and procedures. At stage 6, thinking becomes very abstract. The individual tries to formulate universal, logically consistent principles that can apply to every actor in a moral situation.

Thus, each stage represents a cognitive advance over the prior stage, so it is plausible that children master them in the order that Kohlberg has found.

If Kohlberg's stages do become more differentiated and abstract at each point, the highest stages should be the most difficult to understand. And, in fact, there is some evidence that this is so (Rest, 1973). It also appears that people have a preference for the highest stages they hear, whether they fully understand them or not. People may intuitively sense the greater adequacy of the highest stages, which Kohlberg claims embody the kinds of reasoning which can handle the widest range of moral problems in the most abstract and differentiated way (Kohlberg, 1970, pp. 113-14).

The Problem of Regression

We saw that some subjects appear to skip stages, although this may well be an erroneous finding. It also appears that some subjects, before moving forward, temporarily revert to earlier stages, particularly to stage 2 (Kohlberg and Kramer, 1969). This seems most true of young college students, who temporarily abandon their conventional moral outlooks and enter a phase of relativistic questioning.

This finding is problematic for Kohlberg because he has said that there can be no regressions in his stage sequence (Kohlberg, 1969, p. 99). He therefore argues that these subjects are not really "regressors" but "retrogressors." That is, they represent a special case of stage 4 reasoning, in which subjects can, when asked, still give stage 4 arguments, but they no longer believe in them. Instead, subjects prefer stage 2 arguments. Thus, they are really at a new stage, stage 4B.

Others have objected to Kohlberg's stage 4B solution. For example, Brown and Herrnstein (1975, p. 318) argue that Kohlberg's solution violates the scientific spirit. The scientist must be willing to reject his or her hypotheses when the data disconfirm them. In this case, the data suggested that a kind of regression does occur, so Kohlberg should have simply admitted this fact. Instead, he chose to invent a new stage which enabled him to hold to his earlier position. Admitting that regressions occur, we might add, would not really weaken Kohlberg's theory. Most developmental theories make some room for regression, and Kohlberg probably should do the same.

Movement from Stage to Stage

Many psychologists believe that moral attitudes are the product of *socialization* (see Chapter 13). In this view, parents and other socializing agents gradually transmit, by precept or example, their own moral views to their children.

Kohlberg, a Piagetian, disagrees with the socialization argument. Moral stages do not result from cultural teachings but from children's spontaneous activities. Children develop through social interactions, but interactions of a particular kind. These are interchanges in which others challenge children's assumptions, stimulating them to come up with more comprehensive positions. In such discussions, children do not merely adopt the thoughts of others, but struggle to formulate their own viewpoints (Kohlberg *et al.*, 1975). We will discuss Kohlberg's efforts to promote moral development in a later section.

Stages as Cultural Universals

Socialization theorists (e.g., Bandura, Chapter 13) point out how children acquire different moral beliefs in different cultures. We all, they say, are products of our environment.

Kohlberg does not deny that cultures produce different values, but he still maintains that children progress through his stages in the same order in all cultures. For his stages refer not to specific beliefs, but to underlying modes of reasoning (Kohlberg and Gilligan, 1971). For example, one culture might discourage physical fighting, while another encourages it more. As a result, the children will have different beliefs about fighting, but they still will reason about it in the same way at the same stage. At stage 1, for example, one child might say that it is wrong to fight when insulted, "because you will get punished for it," while another says that "it is all right to fight; you won't get punished." The beliefs differ, but both children reason about them in the same underlying way, in terms of the physical consequences (punishment). They do so because this is what they can cognitively grasp. Later on, the first child might argue that fighting is bad "because if everyone fought all the time there would be anarchy," while the second child argues that "people must defend their honor, because if they don't everyone will be insulting everyone, and the whole society would break down." Once again, the specific beliefs are different, reflecting different cultural teachings, but the underlying reasoning is the same—in this case it is stage 4, where the child can consider something as abstract as the social order. Children, regardless of their beliefs, will always move to stage 4 thinking some time after stage 1 thinking because it is cognitively so much more sophisticated.

Kohlberg, then, proposes that his stage sequence will be the same in all cultures, for each stage is conceptually more advanced than the next. He has given his interview to children in a variety of cultures, including subcultures in Mexico, Taiwan, Turkey, and the Yucatan, and has reported relationships between stage and age which support his theory (1970). To the extent that children move through his stages, they appear to move in order. (These data, however, are cross-sectional, so more solid support for Kohlberg's theory still awaits longitudinal studies.)

What is most important about cultural factors, in Kohlberg's view, is the way in which they stimulate children to move through the stages at different rates. He has found that in simple, isolated villages in Turkey and the Yucatan, children appear to progress through the stages slowly, and many may never reach a stage as high as 4—the modal stage of middle-class adults in urban areas of the United States (Kohlberg and Gilligan, 1971). Urban areas seem to provide more stimulating experiences in terms of education and cultural diversity. As Keniston (1971) says, "Many universities deliberately confront students with contrasting cultures that give allegiance to alien moral concepts and deliberately provoke students to question the unexamined assumptions of their own childhoods and adolescences" (p. 265). Similarly, when young people grow up in cosmopolitan settings, they encounter the diverse views of different social classes and subcultures. Such experiences stimulate thinking, thus promoting moral development.

Cultural experiences, then, can alter the rate and extent of moral development. They do so by stimulating thinking, not by direct teaching. And even though cultural factors can influence the rate of development, they do not alter the stage sequence. To the extent that children progress through the stages, they do so in the same order.

Moral Thought and Moral Behavior

Kohlberg's scale has to do with moral thinking, not moral action. As everyone knows, there are people who can talk at a high moral level, but their behavior is another matter. Consequently, we would not expect perfect correlations between moral judgment and moral action. Still, Kohlberg thinks that there should be some relationship. As a general hypothesis, he proposes that moral behavior is more consistent, predictable, and responsible at the higher stages. At the earlier stages, children base decisions on the fear of punishment (stage 1), personal whims (stage 2), or on the approval of others (stage 3)—all of which vary with the circumstances. At stages 4, 5, and 6, young people base their judgments on laws and/or principles that are relatively stable over time, so they should behave more consistently in a moral fashion (Kohlberg et al., 1975). Some evidence seems to support Kohlberg's view, but good research on this problem is still scarce (see Brown and Herrnstein, 1975, pp. 326-38; Kurtines and Grief, 1974).

Some interesting research has investigated the relationship between moral thinking and disobedience. A crucial political problem, it seems, is people's inability to disobey authority when it asks them to perform inhumane acts. In Nazi Germany, for example, many were merely following orders. The situation has been recreated, to an extent, in laboratory experiments where an authority figure orders the subject to harm someone or implies that the subject should refrain from helping someone in distress. The unsettling finding is that most people do what the person in authority says. There is some evidence, however, that those with some stage 6 moral thinking are more likely to resist, apparently because they are accustomed to thinking in moral terms that go beyond obedience to conventional authority. More research is needed, however, because the findings are not always consistent (Brown and Herrnstein, 1975, p. 332; Podd, 1972; Liebert et al., 1977, p. 374).

Other research has investigated civil disobedience in real life. In particular, researchers have examined the moral thinking of student activists in the 1960s and have found that some activists, in fact, seemed at least partly motivated by postconventional principles (Haan et al., 1968). However, here too the findings are not perfectly consistent, for there were some protests where no postconventional principles were at stake (Keniston, 1971, pp. 260-61).

IMPLICATIONS FOR EDUCATION

Kohlberg would like to see people advance to the highest possible stage of moral thought. The best possible society would contain individuals who not only understand the need for social order (stage 4), but who can entertain visions of universal principles, such as justice and liberty (stage 6) (Kohlberg, 1970).

How, then, can one promote moral development? Turiel (1966) found that when children listened to adults' moral judgments, the resulting change was slight. This is what Kohlberg might have expected, for he believes that if children are to reorganize their thinking, they must be more active.

Accordingly, Kohlberg encouraged another student, Moshe Blatt, to lead discussion groups in which children had a chance to grapple actively with moral issues (Blatt and Kohlberg, 1975). Blatt presented moral dilemmas which engaged the classes in a good deal of heated debate. He tried to leave much of the discussion to the children themselves, stepping in only to summarize, clarify, and sometimes present a view himself (p. 133). He encouraged arguments that were one stage above those of most of the class. In essence, he tried to implement Kohlberg's theory of how children move through the stages. They do so by encountering alternative views which challenge their thinking and stimulate them to formulate better arguments (Kohlberg et al., 1975).

Blatt began a typical discussion by telling a story about a man named Mr. Jones who had a seriously injured son and wanted to rush him to the hospital. Mr. Jones had no car, so he approached a stranger, told him about the situation, and asked to borrow his car. The stranger, however, refused, saying he had an important appointment to keep. So Mr. Jones took the car by force. Blatt then asked whether Mr. Jones should have done that.

In the discussion that followed, one child, Student B, felt that Mr. Jones had a good cause for taking the car and also believed that the stranger could be charged with murder if the son died. Student C pointed out that the stranger violated no law. Student B still felt that the stranger's behavior was somehow wrong, even though he now realized that it was not legally wrong. Thus, Student B was in a kind of conflict. He had a sense of the wrongness of the stranger's behavior, but he could not articulate this sense in terms that would meet the objection. He was challenged to think about the problem more deeply.

In the end, Blatt gave him the answer. The stranger's behavior, Blatt said, was not legally wrong, but morally wrong—wrong according to God's laws (this was a Sunday School class). At this point, Blatt was an authority teaching the "correct" view. In so doing, he might have robbed Student B of the chance to formulate spontaneously his own position. He might have done better to ask a question or to simply clarify the student's conflict (e.g., "So it's not legally wrong, but you still have a sense that it's somehow wrong . . ."). In any case, it seems clear that part of this discussion was valuable for this student. Since he himself struggled to formulate a distinction that could handle an objection, he could fully appreciate and assimilate a new view that he was looking for.

The Kohlberg-Blatt method of inducing cognitive conflict exemplifies Piaget's equilibration model. The child takes one view, becomes confused by discrepant information, and then resolves the confusion by forming a more advanced and comprehensive position. This method also is essentially that of good teaching (Brown and Herrnstein, 1975, p. 321). It is Socratic. The students give a view, the teacher asks questions which get them to see the inadequacies of their views, and they are then motivated to formulate better positions.

Blatt has tried his procedure primarily with children between the ages of 10 and 12 years, giving students between 12 and 18 45-minute sessions. His success has varied. In one class, over half the students advanced a stage; in another, far fewer made this much progress. Blatt probably has achieved only partial success because Socratic teaching is difficult.

One of Blatt's supplementary findings was that those students who reported that they were most "interested" in the discussions made the greatest amount of change. This finding is in keeping with Piagetian theory. Children do not develop because they are shaped through external reinforcements but because their curiosity is aroused. They become interested in information that does not quite fit into their existing cognitive structures and are thereby motivated to revise their thinking.

EVALUATION

Kohlberg, a follower of Piaget, has offered a new, more detailed stage sequence for moral thinking. Whereas Piaget basically found two stages of moral thinking, the second of which emerges in early adolescence, Kohlberg has uncovered additional stages which develop well into adolescence. He has suggested that some people even reach a postconventional level of moral thinking where they think in terms of universal ethical principles which take priority over society's laws and values.

The suggestion of a postconventional morality is unusual in the social sciences. Perhaps it took a cognitive-developmentalist to suggest such a thing. For whereas most social scientists have been impressed by the ways in which societies mold and shape children's thinking, cognitive-developmentalists are more impressed by the capacities for independent thought. If children engage in enough independent thinking, Kohlberg suggests, they will eventually begin to formulate universal principles which they consider higher than social laws and values. This kind of thinking characterizes some of the religious and moral leaders who have at times advocated civil disobedience for the sake of higher ethical considerations.

Many psychologists are skeptical of Kohlberg's claims (see, for example, Kurtines and Grief, 1974). They find it hard to believe that moral reasoning develops in an invariant sequence, regardless of the culture, and find it equally hard to accept Kohlberg's sixth stage. To an extent, the arguments over Kohlberg's research may reflect differences in values; many psychologists may be philosophically opposed to a morality which places itself above the need for social order (see, for example, Hogan, 1975).

In any case, there clearly are weaknesses in the Kohlberg research—or, rather, areas in which more theoretical work and research is necessary. In our discussions, we indicated four main areas. First, there is a need for longitudinal research on the question of invariant sequence. Second, Kohlberg probably needs to be more open to the possibility of regression to earlier stages. Third, we need clearer demonstrations on how Socratic teaching can induce cognitive change. Fourth, more research is needed on the relationships between moral thought and moral behavior. Nevertheless, for the moment we can agree with Brown and Herrnstein (1975, pp. 307-40), who, in a thoughtful review, conclude that the Kohlberg research, despite its limitations, is sufficiently convincing to force us to take his assertions quite seriously.

7

Freud's Psychoanalytic Theory

BIOGRAPHICAL INTRODUCTION

The theorists we have discussed so far, with the exception of the ethologists, have primarily focused on motor or cognitive development. In this chapter, we will begin discussing a group of theorists—the psychoanalysts—whose special province has been the inner world of feelings, impulses, and fantasies. The principal founder of psychoanalytic theory was Sigmund Freud (1856-1939).

Freud was born in Freiberg, Moravia (now a part of Czechoslovakia). He was first child of a 20-year-old mother and a 40-year-old father, although his father also had two grown sons from a previous marriage. The father was a wool merchant who never became very successful in business, and financial troubles forced the family to move twice when Freud was young—first to Leipzig, and then, when Freud was four, to Vienna, where Freud lived until the last year of his life (Jones, 1961, Ch. 1).

As a boy, Freud was a brilliant student, and the family encouraged his studies. His parents made sure that he had an oil lamp to study by, while the other family members had only candles (Schultz, 1975, p. 302). Freud's intellectual interests covered a wide variety of topics, and when he was old enough to enter the university he had difficulty deciding on an area of study. With some reluctance, he chose medicine, primarily because it gave him an opportunity to do research. In medical school, Freud conducted important investigations of the spinal cord of the *Petromyzon,* a type of fish (Jones, 1961, Chs. 3 and 4).

Between the ages of 26 and 35, Freud restlessly searched for a field in which he might make some important discovery. He continued to do research in established areas of neurology, but he was more excited by other possibilities. For a while, he thought he might find revolutionary uses for cocaine, a drug to which he seemed temporarily addicted. Freud also visited Charcot's laboratory in Paris, where Charcot was investigating the mysteries of hysteria. It was the study of this disorder which became the starting point of Freud's great contributions (Jones, 1961, Chs. 5 and 6, 10 and 11).

The term hysteria is applied to physical ailments as well as to losses of memory for which there is no physiological explanation. For example, a woman might complain of a "glove anesthesia," a loss of feeling in the hand up to the wrist, even though physiologically there is no way she could lose sensation in precisely this part of the body.

Freud's first work on hysteria followed the example of Josef Breuer, who had treated a woman ("Anna O.") by helping her uncover buried thoughts and feelings through hypnosis. It seemed to Breuer and Freud (1895) that hysteric patients had somehow blocked off, or *repressed,* wishes and emotions from awareness. The blocked-off energy had then become converted into physical symptoms. Therapy, then, consisted of uncovering and releasing emotions which had been relegated to a separate part of the mind—the *unconscious.*

Freud's early work with hysterics can be illustrated by the case of a woman he called Elizabeth von R (1895). Elizabeth suffered from hysterical pains in her thighs, pains which became worse after walks with her brother-in-law, toward whom she "grew to feel a peculiar sympathy . . . which easily passed with her for family tenderness" (Freud, 1910, p. 23). The sister (his wife) then died, and Elizabeth was summoned to the funeral. As Elizabeth "stood by the bedside of her dead sister, for one short moment there surged up in her mind an idea, which might be framed in these words: 'Now he is free and can marry me' " (p. 23). This wish was totally unacceptable to her sense of morality, so she immediately repressed it. She then fell ill with severe hysterical pains, and when Freud came to treat her, she had completely forgotten the scene at her sister's bedside. Many hours of psychoanalytic work were necessary to uncover this and other memories, for Elizabeth had strong reasons for barring them from consciousness. Eventually, she was able to gain awareness of her feelings, and, to the extent she could accept them, they no longer needed to be redirected into bodily symptoms.

In Freud's work with Elizabeth and many other patients, he did not use hypnosis, the technique Breuer had employed. Freud found that hypnosis, among its other drawbacks, could only be used with some patients, and even with those it often produced only temporary cures. In its place, he developed the method of *free association,* in which the patient is instructed to let his or her mind go and to report everything just as it occurs, making no effort to order or censor the thoughts in any way.

Freud found, however, that although free association eventually leads to buried thoughts and feelings, it is by no means completely free. Patients strongly resist the process. They block on certain topics, change the topic, insist that their thoughts are too trivial or embarrassing to mention, and so on (Freud, 1920, pp. 249-50). Freud named these interruptions *resistance* and considered resistance new evidence for the power of repression in the mind (Breuer and Freud, 1895, p. 314). That is, Freud saw new evidence for his theory that the patient's mind is at war with itself, that certain wishes are unacceptable to the patient's "ethical, aesthetic, or personal pretensions," and that the wishes therefore need to be repressed (Freud, 1910, p. 22).

As Freud built his theory, he speculated that not only hysterics and other neurotic patients suffer from this kind of internal conflict. We all have thoughts and desires that we cannot admit to ourselves. In neurosis, repression and conflict become particularly intense and unmanageable, and symptoms result. Nevertheless, conflict characterizes the human condition (Freud, 1900, p. 294; 1933, p. 121).

Breuer and Freud published a book together—*Studies on Hysteria* (1895)—which became the first classic work in psychoanalytic theory. Afterward, however, Breuer discontinued his work in the area. Breuer's decision was largely influenced by the direction the work was taking. Freud was increasingly finding that the central emotions that hysterics blocked from awareness were sexual ones—a finding which Breuer sensed was true but which he also found personally distasteful and troubling. Moreover, the sexual theory brought ridicule from the scientific community, and this hurt Breuer deeply. Consequently, Breuer left Freud to investigate this new area by himself.

As Freud pressed on with the work, he found that his patients' buried memories led farther and farther back into their pasts—into their childhoods. Freud had great trouble understanding what he was finding. His patients repeatedly told stories about how their parents had committed the most immoral sexual acts against them as children—stories which Freud finally realized must be mere fantasies. For a while, it seemed that his research had gone up in smoke. It was not built on truth, but on fiction. But he then concluded that fantasies, too, govern our lives. Our thoughts and feelings can be as important as actual events (Freud, 1914a, p. 300).

In 1897, the year in which Freud was puzzling over the truth of his patients' memories, he began a second line of investigation—a self-analysis. Motivated by the disturbance he felt when his father died, he began examining his own dreams, memories, and childhood experiences. Through this analysis, he gained independent confirmation of his theory of childhood sexuality and discovered what he considered his greatest insight: the Oedipus complex in the child. That is, he discovered that he (and presumably all children as well) develop an intense rivalry with the parent of the same sex for the affection of the parent of the opposite sex. Freud first published this theory in the *Interpretation of Dreams* (1900). He called the interpretation of dreams "the royal road to the unconscious" (1900, p. 647).

Freud's self-analysis was not an easy process. He had begun delving into an area—the unconscious—out of which "God knows what kind of beast will creep" (Jones, 1961, p. 213). At times, Freud was unable to think or write; he experienced "an intellectual paralysis such as I have never imagined" (Jones, 1961, p. 213). And, on top of this, what he was finding—evidence for childhood sexuality—was unacceptable to most of the scientific community. Most of his colleagues believed, with everyone else, that sexuality begins at puberty, not before. Freud's suggestion that innocent children experience sexual desires indicated that he was little more than a perverted sex maniac. In the face of this reaction, Freud felt "completely isolated" and said that he often dreaded losing his way and his confidence (Freud, 1914a, p. 302).

About 1901 (when Freud was 45 years old), he finally began to emerge from his intellectual isolation. His work attracted various younger scientists and writers, some of whom began meeting with him for weekly discussions. These discussion groups gradually evolved into a formal psychoanalytic association. Among Freud's early disciples were Alfred Adler and Carl Jung, who, like several others, eventually broke with Freud and established their own psychoanalytic theories.

Freud continued to develop and revise his theory until the end of his life, the last 16 years of which he spent in pain from cancer of the jaw. In 1933 the Nazis burned his books in Berlin, and in 1938 he had to leave Vienna for London, where he lived his last year and died at the age of 83.

THE STAGES OF PSYCHOSEXUAL DEVELOPMENT

We have seen how Freud's work led him to believe that sexual feelings must be active in childhood. Freud's concept of sex, however, was very broad. In his view (1905), "sex" includes not just sexual intercourse but practically anything that produces bodily pleasure. In childhood, in particular, sexual feelings are very general and diffuse. Sexual feelings may be included in activities such as sucking for pleasure, masturbation, the wish to show off one's body or to look at the bodies of others, anal excretion or retention, body movements such as rocking, and even acts of cruelty, such as pinching or biting (Freud, 1905, pp. 585-94).

Freud had two major reasons for considering such diverse activities as sexual. First, children seem to derive pleasure from them. For example, babies enjoy sucking even when they are not hungry; they suck their hands, fingers, and other objects because it produces pleasurable sensations on the mucous membranes of the mouth (1905, p. 588). Second, Freud regarded many childhood activities as sexual because they later reemerge in adult sexual activity. For example, most adults engage in sucking (i.e., kissing), looking, exhibitionism, or cuddling immediately prior to and during sexual intercourse. Sometimes, in the cases of so-called perversions, adults reach orgasm through childhood sexual activities alone (without sexual intercourse). For example, a "Peeping Tom" may reach orgasm simply by looking at the bodies of others. Neurotic adults, too, retain childhood sexual wishes, but they feel so much guilt and shame that they repress them (Freud, 1920, Chs. 20 and 21; 1905, pp. 577-79).

In Freud's theory, the term for one's general sexual energy is *libido*, and any part of the body on which this energy becomes focused is called an *erogenous zone* (Freud, 1905, pp. 585-94, 611). Almost any part of the body can become an erogenous zone, but in childhood the three most important zones are the mouth, the anus, and the genital area. These zones become the center of the child's sexual interests in a specific *stage sequence*. The child's first interests center on the mouth (the oral stage), followed by the anus (the anal stage), and finally the genital region (the phallic stage). Freud thought that this sequence is governed by a *maturational* process—by innate, biological factors (1905, p. 587, 621). At the same time, the child's *experiences* also play a decisive developmental role. For example, a child who experiences a great deal of frustration at the oral stage may develop a lasting preoccupation with things having to do with the mouth. Let us now look at Freud's stages in more detail.

The Oral Stage

The first months. Freud said that "if the infant could express itself, it would undoubtedly acknowledge that the act of sucking at its mother's breast is far and away the most important thing in life" (1920, p. 323). Sucking is vital, of course, because it provides nourishment; the baby must suck to stay alive. But, as mentioned, Freud thought that sucking also provides pleasure in its own right. This is why babies suck on their thumbs and other objects even when they are not hungry. Freud called such pleasure-sucking *autoerotic*; when babies suck their thumbs they do not direct their impulses toward others but find gratification through their own bodies (1905, p. 586).

Autoerotic activities are not confined to the oral stage. Later on, for example, children masturbate, and this too is autoerotic. However, Freud emphasized the autoerotic nature of the oral stage because he wanted to stress the extent to which babies are wrapped up in their own bodies. Like Piaget, Freud thought that during the first six months or so of life the baby's world is "objectless." That is, the baby has no conception of people or things existing in their own right. When nursing, for example, young infants experience the comfort of the mother's hold, but they do not recognize the existence of the mother as a separate person. Similarly, when cold, wet, or hungry, babies feel mounting tension and even panic, but they are unaware of any separate person who might relieve the pain. They simply long for a return of pleasurable feelings. Thus, although babies are completely dependent on others, they are unaware of this fact because they do not yet recognize other people's separate existence.

Sometimes Freud described this initial objectless state as one of *primary narcissism* (e.g., 1915a, p. 79). The term narcissism means self-love and is taken from the Greek myth about a boy called Narcissus, who fell in love with his reflection in a pond. As some scholars have observed (e.g., Jacobson, 1954), this term is somewhat confusing because it implies that babies have a clear conception of themselves to love, when they still cannot distinguish themselves from the rest of the world. Still, the term narcissism does convey the idea that at first babies focus primarily inward, on their own bodies. The basic narcissistic state, Freud said (1916), is sleep, when infants feel warm and content and have absolutely no interest in the outside world.

The second part of the oral stage. Beginning at about six months, babies begin to develop a conception of others, especially the mother, as a separate, necessary person. They become anxious when she leaves or when they encounter a stranger in her place (Freud, 1936a, p. 99).

At the same time, another important development is taking place: the growth of teeth and the urge to bite. It is at this point, Karl Abraham (1924a) pointed out, that babies dimly form the idea that it is they, with their urge to bite and devour, who can drive their mothers away. Life at this stage, then, becomes increasingly complex and troubling. It is little wonder that we may often unconsciously wish to return to the earlier oral stage, when things seemed so much simpler and more gratifying.

An illustration: Hansel and Gretel. Freud was aware of the difficulty in reaching conclusions about the infant's mental life. Babies cannot talk and tell us about their feelings and fantasies. To some extent, we are forced to reconstruct the infant's psychic life from the analyses of adults who seem to revert to early ways of thinking, namely psychotics. But Freudians also suggest that many myths and fairy tales reveal the child's early fantasies and concerns.

Bettelheim (1976, pp. 159-66) has written about the oral themes contained in the story of Hansel and Gretel. Briefly, Hansel and Gretel are two children who are sent into the forest by their parents, especially the mother, because they are careless with food (milk) and there is no longer enough to feed them. In the forest, they come upon a gingerbread house, which they proceed to devour. They sense that it may be dangerous to eat so much of the house and hear a voice which asks, "Who is nibbling at my house?" But they ignore it, telling themselves, "It is only the wind" (The Grimms, 1972, p. 90). The woman who owns the house then appears, and she is at first completely gratifying. She gives them all kinds of good things to eat and nice beds in which to sleep. But the next day she turns out to be worse than their mother. She is a witch who intends to eat them.

In Bettelheim's analysis, the themes are largely those of the second oral stage. The story begins with the children experiencing the dreaded separation from their caretakers. There is some hint that the children's own inner urges are at the root of their troubles; they have been reckless with their mother's milk and they greedily devour the gingerbread house. The children's wish is to return to the first oral stage, which seemed so blissful. So they meet the witch, who is temporarily "the original, all-giving mother, whom every child hopes to find again later somewhere out in the world" (Bettelheim, 1976, p. 161). However, this proves impossible. Because they are dimly aware of their own oral destructiveness, they imagine that others will take an oral revenge, which is what the witch attempts to do.

Bettelheim says that fairy tales facilitate growth by addressing children's deepest fears while, at the same time, showing them that their problems have solutions. In this story, Hansel and Gretel finally quit acting solely on the basis of their oral impulses and use more rational parts of the personality. They employ reason to outwit the witch and kill her, and they return home as more mature children.

Fixation and regression. According to Freud, we all go through the oral stage as well as every other stage of psychosexual development. However, we also can develop a *fixation* at any stage, which means that no matter how far we have advanced beyond it, we maintain a lasting preoccupation with the pleasures and issues of the earlier stage. For example, if we are fixated at the oral stage, we might find ourselves continually preoccupied with food; or we find that we work most comfortably when we are sucking or biting on objects, such as pencils; or we gain the most pleasure from oral sexual activities; or we find ourselves addicted to smoking or drinking partly because of the oral pleasure involved (Freud, 1905; Abraham, 1924b).

Freud said that he was not certain about the causes of fixation (1920, p. 357), but psychoanalysts generally believe that fixations are produced by either excessive gratification or excessive frustration at the stage in question (Abraham, 1924b, p. 357; Fenichel, 1945, p. 65). For example, the baby who receives prolonged and very satisfying nursing may continue to seek oral pleasures. Alternatively, the baby who experiences sharp frustrations and deprivations at the oral stage may act as if he or she is unwilling to give up oral satisfactions or as if there is a persistent danger that oral needs will not be met. Such a person might, for instance, become anxious when meals are not served on time and devour the food as if it might disappear at any moment. In general, it seems that severe frustrations, rather than excessive gratifications, produce the strongest fixations (White and Watt, 1974, p. 136, 148, 189; Whiting and Child, 1953).

Sometimes people show few oral traits in their daily lives until they experience some frustration, and then they *regress* to the oral fixation point. For example, a little boy who suddenly finds himself deprived of parental affections when his baby sister is born might regress to oral behavior and once again take up thumb-sucking—something he had previously given up. Or a teenage girl may not be particularly concerned about oral matters until she loses a boyfriend, and then she becomes depressed and finds comfort in eating.

The tendency to regress is determined both by the strength of the fixation in childhood and the magnitude of the current frustration (Freud, 1920, Ch. 22). If we have a strong oral fixation, for example, a relatively small frustration in our current life may be sufficient to cause an oral regression. Alternatively, a major frustration might cause a regression to an earlier developmental stage even if the fixation was not particularly strong.

The kinds of regressions we have been discussing might occur in any of us—in relatively "normal" people. We all find life frustrating at times, and now and then we regress to earlier, more infantile, ways of behaving. Such regressions are not pathological because they are only partial and temporary. For example, the boy who resumes thumb-sucking when his baby sister is born usually does so only for a while; in any case, he does not become like an infant in other respects.

Freud also believed, however, that the concepts of fixation and regression can help clarify more serious emotional disorders. In certain forms of schizophrenia, for example, there is a very complete regression to the first developmental stage. The schizophrenic often withdraws from interaction with others and entertains grandiose ideas concerning his or her importance. The patient may think that he or she is God and that his or her ideas affect the whole world. In such a case, the person has undergone a fairly complete regression to a state of primary narcissism, in which the libido is invested solely in the self, and the boundaries between self and the rest of the world have once again become unstable (Freud, 1920, pp. 422-24).

According to Abraham (1924a), regression to the oral stage is also evident in severe cases of depression. Such depressions frequently follow the loss of a loved one, and a common symptom is the patient's refusal to eat. Perhaps patients are punishing themselves because they unconsciously feel that it was their own oral anger that destroyed the love object.

The Anal Stage

Between the ages of about one and a half and three years, the anal zone becomes the focus of the child's sexual interests. Children become increasingly aware of the pleasurable sensations that bowel movements produce on the mucous membranes of the anal region. As they gain maturational control over their sphincter muscles, they sometimes learn to hold back their bowel movements until the last moment, thereby increasing the pressure on the rectum and heightening the pleasure of the final release (Freud, 1905, p. 589). Children also frequently take an interest in the products of their labors and enjoy handling and smearing their feces (Freud, 1913, pp. 88-91; Jones, 1918, p. 424).

It is at this stage that children are first asked to renounce their instinctual pleasures in a fairly dramatic way. Few parents are willing to permit their children to smear and play with feces for very long. Most parents, as well-socialized individuals, feel a certain repugnance over anal matters and soon get children to feel the same way. As soon as their children are ready, if not sooner, parents toilet-train them.

Some children initially fight back by deliberately soiling themselves (Freud, 1905, p. 591). They also sometimes rebel by becoming wasteful, disorderly, and messy—traits which sometimes persist into adulthood as aspects of the "anal expulsive" character (Hall, 1954, p. 108; Brown, 1940).

Freud, however, was most interested in the opposite reaction to parental demands. He observed that some people develop an excessive stake in cleanliness, orderliness, and reliability (1908a). It seems as if they felt, as children, that it was too risky to rebel against parental demands, and so they anxiously conformed to parental rules instead. Instead of messing and smearing, they became models of self-control, acquiring a disgust for anything dirty or smelly, and developing a compulsive need to be clean and orderly. Such people, who are sometimes labeled "anal compulsive" characters, also harbor resentment over submitting to authority, but they do not dare to express their anger openly. Instead, they frequently develop a passive obstinacy; they insist on doing things according to their own schedule—often while others are forced to wait. They also may be frugal and stingy. It is as if they feel that although they were forced to give up their feces when others demanded it, they will hold onto other things, such as money, and nobody will take them away.

Toilet-training probably arouses sufficient anger and fear to produce some measure of fixation in most children, especially in the United States, where we tend to be strict about this matter (Munroe, 1955, p. 287). Consequently, most people probably develop at least some tendency toward "anal expulsiveness," "anal compulsiveness," or some combination of both. Sometimes these traits have little serious impact on one's life but then emerge in a more pronounced way when one is under stress. For example, writers may be prone to compulsive behavior when they become anxious about their work. A writer may be unable to finish a manuscript because of a compulsive need to check and recheck it for mistakes. To a Freudian, such behavior probably represents a regression to the anal stage, where the individual learned that his or her natural actions met with unexpected disapproval. That is, the writer might have learned that his or her first "productions" were considered dirty and revolting when done spontaneously but prized if done properly. Thus, the writer, anxious about the impact of the manuscript, tries to protect himself or herself by seeing that everything is done precisely as it is supposed to be.

The Phallic or Oedipal Stage

Between the ages of about three and six years, the child enters the phallic or oedipal stage. Freud understood this stage better in the case of the boy than in the case of the girl, so we will begin our discussion with the boy.

The boy's oedipal crisis. The oedipal crisis begins when the boy starts to take an interest in his penis. This organ, which is "so easily excitable and change-able, and so rich in sensations," fires his curiosity (Freud, 1923, p. 246). He wants to compare his penis to those of other males and of animals, and he tries to see the sexual organs of girls and women. He also may enjoy exhibiting his penis and, more generally, imagines the role he might play as an adult, sexual person. He initiates experiments and spins fantasies in which he is an aggressive, heroic male, frequently directing his intentions toward his primary love-object, his mother. He may begin kissing Mommy aggressively, or want to sleep with her at night, or imagine marrying her. He probably does not yet conceive of sexual intercourse *per se*, but he does wonder what he might do with her.

The boy soon learns, however, that his most ambitious experiments and plans are considered excessive and improper. He learns that he cannot, after all, marry Mommy or engage in any sex play with her. He cannot even touch, hug, or cuddle with Mommy as much as he would like, since he is now a "big boy." At the same time, he notices that Daddy seems to be able to do whatever *he* wants; Daddy seems to kiss and hug Mommy at will, and he sleeps with her all night long (doing with her whatever grown-ups do at night). Thus, the lines of the Oedipus complex are drawn: The boy sees the father as a rival for the affections of the mother.

The little boy's oedipal wishes are illustrated by Freud's case of Little Hans (1909). When Hans was about five years, he asked his mother to touch his penis, and he wanted to cuddle with her at night. His father, however, objected. Soon after, Hans had the following dream:

> In the night there was a big giraffe in the room and a crumpled one; and the big one called out because I took the crumpled one away from it. Then it stopped calling out; and then I sat down on top of the crumpled one (p. 179).

According to Freud, Hans's dream probably represented his wish to take his mother (the crumpled giraffe) from the father (the big giraffe).

The little boy, of course, cannot realistically hope to carry out his rivalrous wishes; the father is too big. Of course, he could still entertain rivalrous fantasies, but these too become dangerous. For one thing, he not only feels

FIGURE 7.1   Drawings by a five-year-old boy and a six-year-old girl suggest sexual interests of the phallic stage.

jealous of his father, but loves and needs his father, so he is frightened by his destructive wishes toward him. But more important, the boy begins to consider the possibility of *castration.* In Freud's day, parents often made outright threats of castration when the boy masturbated. Today, parents may discourage masturbation more delicately, but the boy does probably begin to worry about castration when he realizes that his sister or other females are missing a penis. He then concludes that they once had one, but that it was cut off, and the same thing could happen to him. Thus, the oedipal rivalry takes on a new, dangerous dimension, and the boy must escape the whole situation (Freud, 1924, p. 271).

Typically, the boy resolves the oedipal predicament through a series of defensive maneuvers (Freud, 1923; 1924). He fends off his incestuous desires for his mother through *repression*; that is, he buries any sexual feelings toward her deep into his unconscious. He still loves his mother, of course, but he now admits only to a socially acceptable, "sublimated" love—a pure, higher love. The boy overcomes his rivalry with his father by repressing his hostile feelings and by increasing his *identification* with him. Instead of trying to fight the father, he now becomes more like him, and in this way vicariously enjoys the feeling of being a big man. It is as if the boy were to say, "If you can't beat him, join him."

To overcome the oedipal crisis, finally, the child internalizes a *superego*. That is, he adopts his parents' moral prohibitions as his own, and in this way establishes a kind of internal policeman which guards against dangerous impulses and desires. The superego is similar to what we ordinarily call the conscience; it is an inner voice which reprimands us and makes us feel guilty for bad thoughts and actions. Before the child internalizes a superego, he suffers only from external criticism and punishment. Now, however, he can criticize himself, and thus he possesses an inner fortification against forbidden impulses.

The foregoing review suggests the complexity of the Oedipus complex, but it actually is far more complex than we have indicated. The boy's rivalry and love work both ways—he also rivals the mother for the affection of the father (Freud, 1923, pp. 21-24). The situation also is complicated by the presence of siblings, who also become the objects of love and jealousy (Freud, 1920, p. 343), and by other factors, such as the loss of a parent. We cannot begin to go into the limitless variations here, but the interested reader can refer to Fenichel (1945, pp. 91-98).

Typical outcomes. Typically, when the child resolves the Oedipus complex at the age of six years or so, his rivalrous and incestuous wishes are temporarily driven underground. As we shall see, he enters the latency period, during which he is relatively free of these worries. Nevertheless, oedipal feelings continue to exist in the unconscious. They threaten to break into consciousness once again at puberty and exert a strong influence on the life of the adult. Typically, this impact is felt in two central areas: competition and love.

As the adult male enters into competition with other men, he carries with him the dim knowledge of his first forays into this area. The first time he dared to rival a man his masculinity suffered a sharp setback. Consequently, he may be apprehensive about rivaling men again. In the back of his mind, he is still a little boy, wondering if he can really be a big man (Freud, 1914; Fenichel, 1945, p. 391).

The adult also may feel a sense of guilt over competitive urges. The first time he rivaled a man, he wished to do away with his competitor. He repressed these hostile wishes and established a superego to help fend them off, but he still may dimly feel that the wish to become more successful than others is somehow wrong (Freud, 1936b, p. 311).

Oedipal feelings also influence a man's experiences in love. Freud said that the man "seeks above all the memory-image of the mother" (1905, p. 618). However, this desire has its problems. In the early years, it became associated with castration anxiety and guilt. Consequently, men are sometimes impotent with women who evoke too much of the mother's presence. They become sexually inhibited with women who arouse the deep and tender feelings associated with her, and they are most potent with women whom they regard as mere outlets for their physical needs (Freud, 1912).

Freud thought that everyone undergoes an oedipal crisis, so all men have some of these feelings to a certain degree. Severe problems usually stem from excessive fears experienced as a child. Still, oedipal problems are not as serious as those which develop at earlier periods, when the personality is in a more formative stage.

The girl's Oedipus complex. Freud thought that there was an Oedipus complex for the little girl too, but he admitted that "here our material—for some reason we do not understand—becomes far more shadowy and incomplete" (1924, p. 274). Freud's views on this topic, in broad outline, were as follows. He noted (1933, pp. 122-27) that the girl, by the age of five years or so, becomes disappointed in her mother. She feels deprived because her mother no longer gives her the constant love and care that she required as a baby, and, if new babies are born, she resents the attention they receive. Furthermore, she is increasingly irritated by the mother's prohibitions, such as that on masturbation. Finally, and most upsetting, the girl discovers that she does not have a penis—a fact for which she blames the mother, "who sent her into the world so insufficiently equipped" (1925a, p. 193).

The little girl's genital disappointment is illustrated by an anecdote from Ruth Munroe (1955, pp. 217-18), a psychologist who said that she was skeptical about Freud's theory until one day when she observed her four-year-old daughter in the bathtub with her brother. The daughter suddenly exclaimed, "My weewee (penis) is all gone,"—apparently comparing herself with her brother for the first time. Munroe tried to reassure her, but nothing worked, and for some weeks she objected violently even to being called a girl. Thus, this little girl felt what Freud called *penis envy*, the wish to have a penis and to be like a boy (1933, p. 126).

The little girl does, however, recover her feminine pride. This happens when she begins to appreciate the attentions of her father. The father may have not paid any special attention to his daughter when she was in diapers, but now he may begin to admire her cuteness and growing femininity, calling her his little princess and flirting with her in other ways. Thus inspired, she begins to spin romantic fantasies involving herself and her father. At first her thoughts include a vague wish for his penis, but this soon changes into a wish to have a baby and give it to him as a present.

As with the little boy, the little girl discovers that she lacks sole rights to her new love-object. She realizes that she cannot, after all, marry Daddy, nor can she cuddle, hug, or sleep with him as much as she would like. However, the mother seems to be able to do these things, so she becomes the rival for his affections. Freud said that this oedipal situation might be called the *Electra complex* (1940, p. 99).

What most puzzled Freud about the girl's Oedipus complex was the motivation for its resolution. In the case of the little boy, the primary motivation seemed clear: The boy is frightened by the threat of castration. But the little girl cannot fear castration, for she has no penis to lose. Why, then, does she renounce her oedipal wishes at all? In one essay, Freud said that he simply did not know the answer (1925a, p. 196), but his best guess was that the girl resolves the oedipal crisis because she fears the loss of parental love (1933, p. 87). Thus, she does after all repress her incestuous desires, identify with her mother, and institute a superego to check herself against forbidden impulses and wishes.[1] Still, lacking castration anxiety, her motivation to erect strong defenses against oedipal feelings must be weaker, and, as a result, she must develop a weaker superego. Freud knew that this last conclusion would anger the feminists, but this was where his reasoning led, and he argued that women in fact are less rigid about moral issues (1933, p. 129).

Like the boy, then, the little girl entertains and then abandons rivalrous and incestuous fantasies. In some ways, the later consequences of the oedipal experience would seem similar to those for the boy. For example, the girl too may carry within her the dim knowledge that her first attempt at rivaling a woman for a man's love failed, and she may therefore doubt her future prospects. At the same time, though, the girl's oedipal experiences differed from the boy's, so the effects may differ as well. For example, she had less need to resolve the Oedipus crisis, so her oedipal desires may be more open and transparent later in life (Freud, 1923, p. 129). Furthermore, just before she entered into the oedipal rivalry, she experienced a deep disappointment over being female. This feeling, Freud felt, may lead to a "masculinity complex," in which the woman may avoid intimate relationships with men, since these only remind her of her inferior state, and, instead, try to outdo men by becoming very aggressive and assertive (1933, p. 126).

---

[1] As with the boy, the girl's Oedipus complex is exceedingly complex. Rivalries develop with both parents and with siblings as well.

The Latency Stage

With the establishment of strong defenses against oedipal feelings, the child enters the latency period, which lasts from about age six to eleven years. As the name suggests, sexual and aggressive fantasies are now largely latent; they are kept firmly down, in the unconscious. Freud thought that the repression of sexuality at this time is quite sweeping; it includes not only oedipal feelings and memories, but oral and anal ones as well (1905, pp. 580-85). Because dangerous impulses and fantasies are now kept underground, the child is not excessively bothered by them, and the latency period is one of relative calm. The child is now free to redirect his or her energies into concrete, socially accep-table pursuits, such as sports and games and intellectual activities.

Some of Freud's followers have argued that sexual and aggressive fantasies do not disappear at this time as completely as Freud implied (Blos, 1962, pp. 53-54). For example, an eight-year-old boy is still interested in girls' bodies, and he typically discovers the real facts of life at about this age. Nevertheless, most Freudians agree that sexual concerns lose their frightening and overwhelming character. In general, the latency-age child possesses a new composure and self-control.

Puberty (The Genital Stage)

The stability of the latency period, however, does not last. As Erikson says, "It is only a lull before the storm of puberty" (1959, p. 88). At puberty, which begins at about age 11 for girls and age 13 for boys, sexual energy wells up in full adult force and threatens to wreak havoc with the established defenses. Once again, oedipal feelings threaten to break into consciousness, and now the young person is big enough to carry them out in reality (Freud, 1920, p. 345).

Freud said that from puberty onward the individual's great task is "freeing himself from the parents" (1920, p. 345). For the son, this means releasing his tie to the mother and finding a woman of his own. The boy also must resolve his rivalry with his father and free himself of his father's domination of him. For the daughter, the tasks are the same; she too must separate from the parents and establish a life of her own. Freud noted, however, that independence never comes easily (1905, p. 346). Over the years we have built up strong dependencies upon our parents, and it is painful to separate ourselves emotionally from them. For most of us, the goal of genuine independence is never completely attained.

Anna Freud on Adolescence

Although Freud sketched the general tasks of adolescence, he wrote little about the distinctive stresses and behavior patterns of this stage of life. It was his daughter, Anna Freud, who made many of the first contributions to the psycho-analytic study of adolescence.

Anna Freud's starting point is the same as Freud's: The teenager experiences the dangerous resurgence of oedipal feelings. Typically, the young person is most aware of a growing resentment against the parent of the same sex. Incestuous feelings toward the other parent remain more unconscious.

Anna Freud says that when adolescents first experience the welling up of oedipal feelings, their first impulse is to *take flight*. The teenager feels tense and anxious in the presence of the parents and feels safe only when apart from them. Some adolescents actually run away from home at this time, while many others remain in the house "in the attitude of a boarder" (A. Freud, 1958, p. 269). They shut themselves up in their rooms and feel comfortable only when with their peers.

Sometimes adolescents try to escape their parents by developing a blanket *contempt* for them. Instead of admitting any dependence and love, they take an attitude which is exactly the opposite. It is as if they think that they can become free of parental involvement by thinking absolutely nothing of them. Here again, teenagers may fancy themselves suddenly independent, but their parents still dominate their lives, for they spend all their energy attacking and deriding their parents (A. Freud, 1958, p. 270).

Adolescents sometimes attempt to defend themselves against feelings and impulses altogether, regardless of the objects to whom their feelings are attached. One strategy is *asceticism*. That is, the adolescent tries to fend off all physical pleasure. For example, boys or girls might adhere to strict diets, deny themselves the pleasures of attractive clothes, dancing, or music, or anything else fun or frivolous, or try to master their bodies through exhausting physical exercise.

Another defense against impulses is *intellectualization*. The adolescent attempts to transfer the problems of sex and aggression onto an abstract, intellectual plane. He or she might construct elaborate theories on the nature of love and the family, and on freedom and authority. While such theories may be brilliant and original, they are also thinly disguised efforts to grapple with oedipal issues on a purely intellectual level (A. Freud, 1936).

Anna Freud observes that adolescent turmoil and the desperate defenses and strategies of this period are actually normal and to be expected. She does not usually recommend psychotherapy; she thinks that the adolescent should be given time and scope to work out his or her own solution. However, parents may need guidance, for there "are few situations in life which are more difficult to cope with than an adolescent son or daughter during the attempt to liberate themselves" (1958, p. 276).

THE AGENCIES OF THE MIND

We have now reviewed the stages of development. Freud's theory contains many other concepts, and we cannot review them all. However, an introduction to Freud does require a look at one other cluster of concepts, those pertaining to the agencies of the mind. Freud was continually revising his ideas on this topic, but his best known concepts are those of the id, ego, and superego.

The Id

The id is the part of the personality that Freud initially called "the unconscious" (e.g., 1915b). It is the most primitive part of the personality, containing the basic biological reflexes and drives. Freud likened the id to a pit "full of seething excitations," all pressing for discharge (1933, p. 73). In terms of motivation, the id is dominated by the *pleasure principle*; its goal is to maximize pleasure and minimize pain. Pleasure, in Freud's view, is primarily a matter of reducing tension (1920, p. 365). For example, during sexual intercourse, tension mounts and its final release is pleasurable. Similarly, we find that the release of hunger or bladder tensions brings pleasurable relief. In general, the id tries to remove all excitation and to return to a quiet state—namely, that of deep, peaceful sleep.

At first, the baby is almost all id. Babies worry about little besides bodily comfort, and they try to discharge all tensions as quickly as possible. However, even babies must experience frustration. For example, they sometimes must wait to be fed. What the id does then is to *hallucinate* an image of the desired object, and in this way it temporarily satisfies itself. We see such *wish-fulfilling* fantasies at work when a starving person hallucinates an image of food, or when a thirsty dreamer dreams that a glass of water is at hand and therefore does not have to wake up and get one (Freud, 1900, p. 158, 165). Such fantasies are prime examples of what Freud called *primary process thinking* (1900, p. 535).

In the course of life, many impressions and impulses are repressed into the id, where they exist side by side with the basic drives. In this "dark and inaccessible" region of the mind, there is nothing that corresponds to logic or a sense of time (Freud, 1933, pp. 73-74). Impressions and strivings "are virtually immortal; after the passage of decades they behave as though they had just occurred" (1933, p. 74). Images in the id, furthermore, are very fluid and easily merge into one another. The id is oceanic, chaotic, and illogical. It is completely cut off from the external world. Our best knowledge of this mysterious region comes from the study of dreams.

The id, then, contains basic drives and reflexes, along with images and sensations that have been repressed. So far we have focused on the id's sexual drives and those associated with the preservation of life, such as hunger and thirst. The id also contains aggressive and destructive forces. Freud's views on aggression are complex and underwent drastic revisions, but in a sense we can see how aggression follows the id's basic principle of reducing tension. In the id, any image that is associated with pain or tension should be instantly destroyed. It does not matter to the id that one may be wishing for the destruction of someone one needs and loves; contradictions such as these have no importance in this illogical region of the mind. The id simply wants a reduction in disturbing tensions immediately.

The Ego

If we were ruled by the id, we would not live for long. To survive, one cannot act solely on the basis of hallucinations nor simply follow one's impulses. We must learn to deal with reality. For example, a little boy soon learns that he cannot just impulsively grab food from wherever he sees it. If he takes it from a bigger boy, he is likely to get hit. He must learn to consider reality before acting. The agency that delays the immediate impulse and considers reality is called the ego.

Freud said that whereas "the id stands for the untamed passions," the ego "stands for reason and good sense" (1933, p. 76). Because the ego considers reality, it is said to follow the *reality principle* (1911). The ego tries to forestall action until it has had a chance to perceive reality accurately, to consider what has happened in similar situations in the past, and to make realistic plans for the future (1940, p. 15). Such reasonable ways of thinking are called *secondary process thinking* and include what we generally think of as perceptual or cognitive processes. When we work on a math problem, plan a trip, or write an essay, we are making good use of ego functions. At first, though, the ego's functioning is largely bodily or motoric. For example, when a child first learns to walk, he or she inhibits impulses toward random movement, considers where he or she is headed to avoid collisions, and otherwise exercises ego controls (1940, p. 15).

The term "ego" is one which we hear a great deal in everyday language. Quite often, we hear that someone has a "big ego," meaning that this person has an inflated self-image. Although Freud himself occasionally wrote about the ego in just this way (e.g., 1917), many Freudians (e.g., Hartmann, 1956; Beres, 1971) contend that we should distinguish between the ego and the self-image. Strictly speaking, they say, the ego refers only to a set of functions—e.g., judging reality accurately, regulating impulses, and so on. The self-image, the picture we have of ourselves, is different from the ego itself.

Freud emphasized that although the ego functions somewhat independently from the id, it also borrows all of its energy from the id. He likened the ego's relation to the id to that of a rider on a horse. "The horse supplies the locomotive energy, while the rider has the privilege of deciding on the goal and guiding the powerful animal's movement. But only too often there arises between the ego and the id the not precisely ideal situation of the rider being obliged to guide the horse along the path by which it itself wants to go" (1933, p. 77).

The Superego

The ego is sometimes called one of the "control systems" of the personality (Redl and Wineman, 1951). The ego controls the blind passions of the id to protect the organism from injury. For example, we mentioned how a little boy must learn to inhibit the impulse to grab food until he can determine whether it is realistically safe to do so. But we also control our actions for other reasons. We might also refrain from taking things from others because we believe such actions are morally wrong. Our standards of right and wrong constitute the second control system of the personality, the superego.

We referred earlier to Freud's view on the origin of the superego: It is a product of the oedipal crisis. Children introject parental standards to check themselves against the dangerous impulses and fantasies of this period. Freud did note, though, that the superego continues to develop after this period as well. Children continue to identify with other people, such as teachers and religious leaders, and to adopt their moral standards as their own (1923, p. 27).

Freud wrote about the superego as if it contains two parts (1923, pp. 24-25). One part is sometimes called the *conscience* (Hall, 1954). It is the punitive, negative, and critical part of the superego which tells us what *not* to do and punishes us with feelings of guilt when we violate its demands. The other part is called the *ego ideal,* and this part consists of *positive* aspirations. For example, when a little boy wants to be just like a famous basketball player, the athlete is his ego ideal. The ego ideal also may be more abstract. It may include our positive ideals, such as the wish to become more generous, courageous, or dedicated to principles of justice and freedom.

Levels of Awareness of the Three Agencies

The id, ego, and superego function at differing levels of awareness, as Freud tried to show by means of a diagram (1933, p. 78). The diagram is reproduced here as Figure 7.2.

The id, at the bottom of the drawing, is completely removed from the region labeled "pcpt.-cs," from consciousness and the perception of reality. The id is entirely unconscious, which means that its workings can be made conscious only with a great deal of effort.

The ego extends into consciousness and reality; it is the part of the id which develops in order to deal with the external world. The ego, you will note, largely inhabits a region labeled "preconscious." This term refers to functioning which is below awareness but which can be made conscious with relatively little effort. The ego also is partly unconscious; for example, it represses forbidden thoughts in a completely unconscious way.

The superego is drawn on top of the ego, illustrating its role of criticizing the ego from above. The superego, too, is partly unconscious; although we are sometimes aware of our moral standards, they also frequently affect us unconsciously. For example, we might suddenly become depressed without any idea why, because our superego is punishing us for forbidden thoughts.

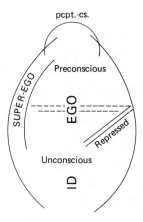

FIGURE 7.2   Freud's sketch of the personality structure.

Reprinted from *New Introductory Lectures on Psycho-analysis* by Sigmund Freud. Translated and edited by James Strachey. Copyright 1933 by Sigmund Freud. Copyright renewed 1961 by W. J. H. Sprott. Copyright © 1965, 1964 by James Strachey. By permission of W. W. Norton & Co., Inc., The Hogarth Press, and George Allen & Unwin Ltd.

The Central Role of the Ego

In the last analysis, our ability to deal with life—our mental balance—rests with our ego's ability to meet the various pressures put upon it. More specifically, the ego is the executive agency which must somehow serve three "tyrannical masters"—the id, reality, and the superego (1933, p. 78). The ego must eventually meet the biological demands of the id, but in a way that also respects external reality and does not violate the precepts of the superego. The task is difficult because the ego is basically weak; as mentioned, it has no energy of its own, but borrows it from the id. Consequently, the ego is quite prone to anxiety —to the fear that it might not satisfy any one of its three dictators. First, the ego experiences "neurotic anxiety" when it feels it cannot master the impulses of the id, as when a dangerous, repressed wish threatens to break out, or when we feel helpless to meet a basic biological need. Second, the ego feels "moral anxiety" when it anticipates punishment from the superego. And third, the ego is always subject to "realistic anxiety," to the dangers in the external world.

Thus, "the ego, driven by the id, confined by the superego, repulsed by reality, struggles to master its economic task of bringing harmony among the forces and influences working in and upon it; and we can understand how it is that so often we cannot suppress a cry: 'Life is not easy!' " (Freud, 1933, p. 78).

Dreams

The three agencies of the mind are all at work during dreaming. Dreams begin as wishes from the id. For example, a child who is hungry dreams that he or she is eating a delicious meal. Such dreams are innocent enough and directly express the id's wishes. However, many wishes violate the standards of the superego, so the ego requires some distortion or disguise of the original wish before permitting the dream to surface into consciousness (Freud, 1940, p. 51). In one of Freud's examples, a little boy was told that it was wrong for him to eat very much because of his illness; so one night, when he was especially hungry, he dreamt that someone else was enjoying a sumptuous meal (1900, pp. 301-2). Similarly, many sexual and aggressive wishes must be disguised and distorted before surfacing. Because the dream we remember upon waking has undergone disguise, we cannot interpret it in any simple, straightforward way. In psychoanalysis, the patient is asked to free associate to the dream. That is, he or she is asked to say whatever comes to mind in connection with each aspect of it. Patients do not find it easy to associate freely—without censoring "immoral" thoughts and fantasies—but eventually the process can reveal the dream's underlying meanings (Freud, 1940, p. 53; 1920, pp. 249-50).

The Ego's Development

In their elaborations on his theories, Freud's followers have been primarily concerned with the role of the ego. In 1936 Anna Freud pointed out that although the ego may be weak, it is the ego that employs various mechanisms of defense, such as repression and intellectualization. The healthiest defense mechanism is sublimation, by which the ego desexualizes energy and channels it into socially acceptable pursuits (S. Freud, 1920, p. 354).

Freudians also have been concerned with the process by which the ego develops. Freud wrote little on this subject. His main suggestion (1911) was that the ego develops because the drives are frustrated. As long as babies are gratified, they have no reason for dealing with reality. But they do experience frustration. At first they try to reduce tensions through hallucinations, but these do not work for long. So they must seek need-gratifying objects in reality.

The difficulty with Freud's proposal is that it implies that the ego only acts when the id activates it. Construed in this way, the ego is weak. It only serves the id.

Hartmann's revision. One of Freud's most influential followers, Heinz Hartmann, suggested that Freud's theory might permit a different picture of the ego—one that gives it more autonomy. Hartmann (1939, 1950) noted that Freud wondered whether the ego might not have genetic roots of its own. If so, ego functions such as motility (body movement), language, perception, and cognition might develop according to their own *maturational timetable*. Children, then, might begin to walk, talk, grasp objects, and so on from inner promptings that are biologically governed, but which also are independent of the instinctual drives. Children have a maturational need to develop ego functions when the id is at rest, when life is "conflict-free." Thus, the ego might develop independently from the id. Hartmann's proposal is widely considered a major breakthrough in the study of ego development.

Ego development and object relations. Freudians also have studied the kind of environment most conducive to ego growth. When Freudians talk about the environment, they usually are speaking not about the physical world but the world of other people. They call interactions with others "object relations."

According to Hartmann (Hartmann *et al.*, 1946), Benedek (1938), and others, the optimal interpersonal environment at first is a consistently gratifying one. When caretakers are consistently loving, babies become interested in them and learn about the external world of people. Consistent care seems especially important for the development of one essential ego function—the ability to postpone gratification. As Benedek says, when the mother is consistent and loving, the baby gains *confidence* in her. The baby does not grow unduly impatient the minute his or her needs are not met but learns to wait. The baby knows that care is forthcoming.

When, in contrast, babies suffer painful deprivations and inconsistent care, they often have great difficulty postponing gratification. They may become overly aggressive and impulsive, feeling that if they do not take something immediately, it may be missing when they need it (Redl and Wineman, 1951).

PRACTICAL IMPLICATIONS

It is impossible to make any sharp distinctions between Freudian theory and practice. Freud's theory emerged from the clinical treatment of patients, and his followers continue to use therapy as a main source of data. In this chapter, we have focused on Freud's theory of normal growth and development, rather than pathology and treatment, but we still have found it necessary to mention topics in the latter areas (e.g., hysteria).

A description of Freud's therapeutic work is well beyond our scope. What we can say here is that a major goal of psychoanalysis is to recover repressed or blocked-off experience. For example, we mentioned how this was necessary in the case of Elizabeth von R. Elizabeth had repressed sexual feelings toward her brother-in-law, and these feelings, far from disappearing, became diverted into painful bodily symptoms. The only solution, Freud thought, is for us to become more conscious of our thoughts and feelings so that, instead of their controlling us, we can gain a measure of control over them. As he put it, "Where id was, there ego shall be" (1933, p. 80).

Therapy with children usually proceeds somewhat differently from that with adults, for children are not given to verbal discussions and recollections. Instead, they learn to express, accept, and master feelings and fantasies through play. In the Bettelheim chapter, we will discuss a therapeutic approach with extremely disturbed children.

It is important to note that Freud never believed that psychoanalysis can completely cure our problems. Because we all live in society, which demands some repression of our instinctual urges, we all suffer to some extent. Further, Freud saw the therapist's role as limited. A psychiatrist was once asked by Freud if he was really able to cure. The psychiatrist replied: "In no way, but one can, as a gardener does, remove some impediments to personal growth." "Then," Freud answered, "we will understand each other" (Ellenberger, 1970, p. 461).

The practical implications of Freud's ideas extend far beyond the treatment of patients. His ideas have influenced practically every area of life, including the practice of law, art, literature, religion, and education. The area of most interest to us here is education. Freud's thoughts on education were not as radical as it is sometimes supposed. He believed that societies will always exact some instinctual renunciation, and he said it would be unfair to send children into the world expecting that they can do just as they please (1933, p. 149). On the other hand, Freud thought that discipline is usually excessive; it makes children feel unnecessarily ashamed and guilty about their bodies and their natural functions. Freud was particularly emphatic on the need for sex education (1907). He recommended that sex education be handled by the schools, where children could learn about reproduction in their lessons on nature and animals. They themselves would then draw the necessary conclusions with respect to humans.

Freudian ideas have motivated some more adventurous experiments in education. For example, at Summerhill, A. S. Neill (1960) has given children a great deal of liberty of all kinds, including sexual freedom. However, such radical innovations are rare, and Freud's influence is more typically found in the general attitude a teacher takes toward children. This attitude is particularly evident when a teacher refrains from automatically disciplining some unwanted behavior and instead tries to understand the emotional reasons behind it (Russell, 1971). When a teacher takes a closer look at a child's life, the teacher may discover that the angry or sullen child is not really mad at the teacher but is finding something at home, such as the neglect of a parent, deeply frustrating. Or the teacher may find that the shy child has for many years been made to feel inferior in the presence of adults. Or the teacher may discover that the seemingly lazy teenager is actually brooding endlessly over sex or social failure (White and Watt, 1973, p. 34). The teacher may not always be able to correct such problems, or even feel that it is prudent to discuss them with the student; for the student may need his or her privacy in certain matters. Nevertheless, a measure of understanding can help the teacher. For the teacher is then not so quick to criticize or punish but has a reason for being patient and encouraging—attitudes which have helped many a child.

EVALUATION

Freud was one of the great thinkers of this century. Before Freud, some poets, artists, and philosophers may have known about the unconscious and early sexual fantasies, but it was Freud's remarkable work that made us take these matters seriously. At the same time, Freud was more bitterly attacked than any other psychological theorist before or since. Even today many consider his ideas scandalous.

It is not surprising, then, that Freud and his followers have sometimes reacted dogmatically and defensively. At times they have behaved like members of a religious sect, isolating themselves from other scientists and gathering in their own groups to reaffirm the truth of their own beliefs. At other times, Freudians have resorted to *ad hominum* arguments—arguments directed not against others' ideas, but against their personalities. In one essay (1925b), Freud argued that his critics objected to his ideas because of their own resistances and repressions.

In the midst of such emotionally charged debates, there have been several criticisms of psychoanalysis which have merit, and some Freudians have tried to face them openly and to correct the weaknesses in the theory.

Some of the most important criticisms of Freud have come from anthropologists, who have argued that Freud's theory is culture-bound. In the 1920s Malinowski and others zeroed in on Freud's theory of the Oedipus complex, pointing out that it is not nearly as universal as Freud imagined. Malinowski noted that the family constellation on which this complex is based—the nuclear triangle of mother, father, and child—is not found in all cultures. Among the Trobriand Islanders, Malinowski found, the child's chief disciplinarian was not the father but the maternal uncle. Further, the strongest incest taboo was not between children and parents but between brothers and sisters. In this situation, Malinowski pointed out, repressed fears and longings were very different. "We might say that in the Oedipus complex there is a repressed desire to kill the father and marry the mother, while in the ... Trobriands the wish is to marry the sister and to kill the maternal uncle" (1927, pp. 80-81). Thus, the oedipal situation is by no means just as Freud described.

Malinowski, however, did not wish to dispense with Freud altogether. On the contrary, he was indebted to Freud for the insight that repressed wishes emerge in projections such as dreams, magic, and folklore. This insight provided him with an important theoretical tool. Malinowski's argument was that such projections vary with the cultural setting. Among the Trobriand Islanders, he found no oedipal myths or dreams, but many which centered on their own strongest temptations and taboos—especially brother-sister relations. For example, although they themselves never admitted to incestuous wishes toward siblings, they told stories about how magic originated long ago when a brother and sister did commit incest.

At the time of Malinowski's writings, Freud and his followers resisted anthropological modifications of psychoanalytic theory. More recently, however, several Freudians (e.g., Kardiner, 1945; Erikson, 1950; A. Parsons, 1964) have tried to combine psychoanalytic and anthropological insights.

Freud also has been sharply criticized for cultural bias on the topic of women. Psychoanalytically oriented writers such as Clara Thompson (1950) and modern feminists have charged that Freud's views on women reflect his own unexamined Victorian attitudes. Freud's limitations, Thompson says, are most evident in his concept of penis envy. She agrees that girls envy boys, but not for the reasons Freud thought. Freud assumed that penis envy is based on a real biological inferiority—a view that fit well with his society's prejudice. Actually, she says, penis envy is much more of a cultural problem; girls feel inferior to boys because girls lack the same privileges in a male-dominated society. That is, they lack the opportunities for adventure, independence, and success. Freud ignored women's legitimate desire for social equality.

Writers have also accused Freud of cultural bias in his discussions of women's sense of morality. Freud thought that girls, not fearing castration, have less need to internalize a strong superego. He then pointed to women's greater emotionality and flexibility in moral matters as evidence. Such observations, his critics contend, simply reflect his own cultural stereotypes.

Recent empirical evidence does suggest that Freud would have done well to question his theory on superego formation. Most evidence suggests that children do not acquire an initial sense of morality because they fear harm, whether castration or some other physical punishment. The child who only fears physical punishment simply tries to avoid getting caught (and perhaps learns to hate the punisher). A sense of morality, instead, appears to develop when the child experiences love and wishes to keep it. The child who receives love tries to behave properly to gain parental approval (Brown, 1965, pp. 381-94; White and Watt, 1973, p. 319). Thus, if a girl is loved as much as a boy, she should develop an equally strong conscience.

Freud's theory of the girl's Oedipus complex, then, has met with justifiable criticisms. Still, thoughtful critics such as Clara Thompson do not wish to overturn Freud altogether. The task is to sort out what is valid in the theory. Girls still may develop sexual longings, fears, and rivalries during the oedipal period, and it is important not to overlook these developments.

Freud also has been criticized on scientific grounds. Although his theory hinges on universal childhood developments, his evidence came primarily from adults—from the memories and fantasies of adults in treatment. Freud did not investigate his hypotheses in an unbiased way with representative samples of normal children.

Some psychologists think that Freud's theory is of little scientific value because it is so hopelessly opaque and complex. As Schultz puts it: "It seems that, except in the broadest sense of the word, there is no such thing as a psychoanalytic theory! There are a large number of generalizations and hypotheses, but there seems to be no orderly framework of theorems, postulates, or precise relationships so necessary to a scientific theory" (1975, p. 322). Sometimes, in fact, it seems as if the theory predicts equally probable but contradictory outcomes. For example, children who experience frustration at the anal stage might develop habits of orderliness, cleanliness, and obedience, or they might develop the opposite characteristics, rebelliousness and messiness. How does one predict which set of traits any given child will develop?

Finally, there is the unnerving experience of never seeming able to disconfirm Freud's hypotheses. If one does a study and finds no relationship between weaning and later oral behavior, for example, some Freudians will argue that one failed to understand Freud's thoughts in sufficient depth.

Nevertheless, despite all the problems that people have encountered in researching Freud's ideas, an enormous amount of research has been done and will continue, and investigators will eventually sort out the valid and invalid propositions. Researchers will continue to struggle with Freud's theory and to test it the best they can because they sense that Freud was basically on the right track. As Hall and Lindzey say (1975), his theory has a fundamental appeal because it is both broad and deep.

> Over and above all the other virtues of his theory stands this one—it tries to envisage a full-bodied individual living partly in a world of reality and partly in a world of make-believe, beset by conflicts and inner contradictions, yet capable of rational thought and action, moved by forces of which he has little knowledge and by aspirations which are beyond his reach, by turn confused and clear-headed, frustrated and satisfied, hopeful and despairing, selfish and altruistic; in short, a complex human being. For many people, this picture of man has an essential validity (p. 72).

Erikson
and the Eight
Stages of Life

BIOGRAPHICAL INTRODUCTION

Among the advances in the psychoanalytic theory of development, none have been more substantial than those made by Erik H. Erikson. Erikson has given us a new, enlarged picture of the child's tasks at each of Freud's stages. And, beyond this, he has added three new stages—those of the adult years—so the theory now encompasses the entire life cycle.

Erikson was born in 1902 in Frankfurt, Germany. Actually, his parents were Danish, but they separated a few months before he was born, so his mother went to Frankfurt to have her baby among friends. There she raised Erikson by herself until he was three years old, when she married a local pediatrician, Dr. Homburger. Erikson's mother and stepfather were Jewish, but Erikson looked different—more like a tall, blond, blue-eyed Dane. He was even nicknamed "the goy" (non-Jew) by the Jewish boys (Coles, 1970, p. 180).

Young Erikson was not a particularly good student. Although he excelled in certain subjects—especially ancient history and art—he disliked the formal school atmosphere. When he graduated from high school, he felt lost and uncertain about his future place in life. Instead of going to college, he wandered throughout Europe for a year, returned home to study art for a while, and then set out on his travels once again. He was going through what he would later call a moratorium, a period during which young people take time out to try to find themselves. Such behavior was acceptable for many German youth at the time. As Erikson's biographer, Robert Coles (1970), says, Erikson "was not seen by his family or friends as odd or 'sick,' but as a wandering artist who was trying to come to grips with himself" (p. 15).

Erikson finally began to find his calling when, at the age of 25, he accepted an invitation to teach children in a new Viennese school founded by Anna Freud and Dorothy Burlingham. When Erikson wasn't teaching, he studied child psychoanalysis with Anna Freud and others, and was himself analyzed by her.

At the age of 27 Erikson married Joan Serson and started a family. Their life was disrupted in 1933 when the rise of Hitler forced them to leave Europe. They settled in Boston, where Erikson became the city's first child analyst.

Nevertheless, the urge to travel seemed firmly implanted in Erikson's nature. After three years in Boston, he took a position at Yale, and two years later he made another trip—to the Pine Ridge Reservation in South Dakota where he lived with and learned about the Sioux. Erikson then moved on to San Francisco, where he resumed his clinical practice with children and participated in a major longitudinal study of normal children at the University of California. He also found time to travel up the California coast to study another Indian tribe, the Yurok fishermen. We can see that Erikson was exploring areas that Freud had left uncharted—the lives of normal children and children growing up in different cultural contexts.

In 1949, during the McCarthy era, Erikson came into conflict with his employer, the University of California. The University demanded a loyalty oath of all its employees, which Erikson refused to sign. When some of his colleagues were dismissed, he resigned. Erikson then took a new job at the Austin Riggs Center in Stockbridge, Massachusetts, where he worked until 1960. He was then given a professorship at Harvard, even though he had never earned a formal college degree. He has been at Harvard since then.

Erikson's most important work is *Childhood and Society* (1950). In this book he maps out his eight stages of life and illustrates how these stages are played out in different ways in different cultures. Two other highly influential books are *Young Man Luther* (1958) and *Gandhi's Truth* (1969), which bridge psychoanalytic insights with historical material.

ERIKSON'S STAGE THEORY

General Purpose

Freud, you will recall, postulated a sequence of psychosexual stages which center on body zones. As children mature, their sexual interest shifts from the oral to the anal to the phallic zone; and then, after a latency period, the focus is once again on the genital region. Thus, Freud presented a completely new way of looking at development.

At the same time, however, Freud's stage theory is limited. In particular, its focus on body zones is too specific. As you will recall from Chapter 5, a rigorous stage theory describes *general* achievements or issues at different periods of life. For example, we do not call shoe-tying a stage because it is too specific. Similarly, the focus on zones also tends to be specific, describing only parts of the body. While it is interesting to note that some people become fixated on these zones—and, for example, find the mouth or the anus the main source of pleasure in life—there is more to personality development than this.

Freud's writings, of course, were not limited to descriptions of body zones. He also discussed crucial interactions between children and significant others. Erikson has tried to do this more thoroughly. At each Freudian stage, he has introduced concepts which gradually lead to an understanding of the most decisive, general encounter between the child and the social world.

1. The Oral Stage

Zones and modes. Erikson first tries to give the Freudian stages greater generality by pointing out that for each libidinal zone we can also speak of an ego mode. At the first stage, the primary zone is the mouth, but this zone also possesses a mode of activity, *incorporation,* a passive yet eager taking in (Erikson, 1950, p. 72). Further, incorporation extends beyond the mouth and characterizes other senses as well. Babies not only take in through the mouth, but through the eyes; when they see something interesting, they open their eyes eagerly and widely and try to take the object in with all their might. Also, they seem to take in good feelings through their tactile senses. And even a basic reflex, the grasp reflex, seems to follow the incorporative mode; when an object touches the baby's palm, the fist automatically closes around it. Thus, incorporation describes a general mode through which the baby's ego first deals with the external world.

Freud's second oral stage is marked by the eruption of teeth and aggressive biting. According to Erikson, the mode of *biting* or *grasping,* like incorporation, is a general one that extends beyond the mouth. With maturation, babies can actively reach out and grasp things with their hands. Similarly, "the eyes, first part of a relatively passive system of accepting impressions as they come along, have now learned to focus, to isolate, to 'grasp' objects from a vaguer background, and to follow them" (Erikson, 1950, p. 77). Finally, the organs of hearing conform to the more active mode of grasping. Babies can now discern and localize significant sounds and can move their heads and bodies so as to actively take them in. Thus, the mode of biting or grasping—of active taking—is a general one which describes the central way in which the ego now deals with the world.

The most general stage: Basic Trust vs. Mistrust. The most general stage at each period consists of a general encounter between the child's maturing ego and the social world. At the first stage, as babies try to take in the things they need, they interact with caretakers, who follow their own culture's ways of giving to them. What is most important in these interactions is that babies come to find some consistency, predictability, and reliability in their caretakers' actions. When they sense that a parent is consistent and dependable, they develop a sense of basic trust in the parent. They come to sense that when they are cold, wet, or hungry, they can count on others to relieve their pain. Some parents come promptly, while others minister on schedules, but in either case babies learn that the parent is dependable and therefore trustworthy. The alternative is a sense of mistrust, the feeling that the parent is unpredictable and unreliable, and may not be there when needed (Erikson, 1950, p. 247).

Babies must also learn to trust themselves. This problem becomes particularly acute when babies experience the rages of teething and hurt the nursing mother with their sharp bites and grasps. When babies learn to regulate their urges—to suck without biting and to hold without hurting—they begin to consider themselves "trustworthy enough so that the providers will not need to be on guard lest they be nipped" (1950, p. 248). For her part, the mother needs to be careful not to withdraw too completely or to wean too suddenly. If she does, the baby will feel that her care is not dependable after all, for it may be suddenly taken away.

When babies have developed a sense of trust in their caretakers, they show it in their behavior. Erikson says the first sign of trust in a mother comes when the baby is willing "to let her out of sight without undue anxiety or rage" (1950, p. 47). The word "undue" is probably important here, for we saw in the discussion of Bowlby that most babies experience some separation anxiety. Nevertheless, if parents are dependable, babies can better tolerate their absences. If caretakers are undependable, babies cannot afford to let them go and panic when they begin to do so.

So far, I may have implied that babies should develop trust but not mistrust. Erikson, however, does not mean this. All babies experience both attitudes, and mistrust is itself necessary for growth. "It is clear," Erikson says, "that the human infant must experience a goodly measure of mistrust in order to trust discerningly . . ." (1976, p. 23). What is vital for healthy development is a favorable *ratio* between trust and mistrust; the former should outweigh the latter. The same principle holds for the other stages, which, as we shall see, also are phrased in terms of polarities.

Trust is similar to what Benedek calls confidence (see p. 141 in the preceding chapter). It is a basic faith in one's providers. Trust is an *ego* strength because it enables the child to postpone gratification—one of the ego's central capacities. If a baby boy feels that his mother is reliable, he does not need her attention the instant he feels the pressure from a mounting need. He has learned that she will come in due time. He has learned to wait.

Trust, then, is the sense that others are reliable and predictable. At the same time, however, Erikson implies that trust ultimately depends on something more. Ultimately, he says, trust depends on the *parents' own confidence,* on their sense that they are doing things right. Parents "must be able to represent to the child a deep, an almost somatic conviction that there is a meaning to what they are doing" (1950, p. 249). This sense of meaning, in turn, requires cultural backing—the belief that the "way we do things is good for our children."

At first glance, Erikson's emphasis on the caretaker's own confidence is puzzling. What does the parent's confidence have to do with the baby? Erikson might have in mind thoughts similar to those of the psychiatrist H. S. Sullivan. Sullivan (1953, pp. 49-109) believed that in the first months of life the infant has a special kind of physical empathy with the mother-figure such that the baby automatically feels the mother's state of tension. If the mother feels anxious, the baby feels anxious; if the mother feels calm, the baby feels calm. These early interactions, in turn, influence later attitudes. Thus, it is important that parents feel reasonably confident and self-assured, so babies will not become too wary of interpersonal contact. Babies need to feel that it is basically good and reassuring to be close to others.

Erikson (1959, p. 64) observes that it is not always easy for American parents to have an inner confidence in their child-rearing practices. Whereas parents in simpler, more stable cultures follow practices that have been handed down over the generations, the modern American parent has fewer traditions to fall back on. Modern parents receive all kinds of advice on newer, "better" child-rearing techniques, and the advice is by no means uniform.

In this situation, Erikson thinks that books such as Spock's (1945) are helpful. Throughout his book, Spock encourages parents to trust themselves. He tells parents that they know more than they think and that they should follow their impulses to respond to their babies' needs. It is almost as if Spock had read Erikson and understands the importance of the parent possessing an inner assurance.

Beyond reading Spock, parents can gain an inner security from religion. Their own faith and inner assurance will be transmitted to the child, helping the child feel that the world is a trustworthy place. If parents are without religion, they must find faith in some other area, perhaps in fellowship, or the goals of their work, or in their social ideals (Erikson, 1959, pp. 64-65).

Summary. We see, then, how Erikson has broadened Freud's description of the oral stage. He first shows that it is not just the oral zone that is important but the oral modes of interacting with the world. The first dominant mode is incorporation, which includes not only a tendency to take in through the mouth but through all the senses. Incorporation and, later, biting and grasping, seem to cover general ways the ego interacts with the world. The child's ego, in turn, meets the social world—in this case, the caretakers—in a general, decisive encounter. The crucial issue is the extent to which the baby can develop a sense of trust. This trust depends on the caretakers' predictability, as well as their trust in themselves. Caretakers gain confidence from their faith in some aspects of the culture, which may include the simple advice to go ahead and follow their inclinations to meet the baby's biological needs. Thus we see how Erikson's thinking extends beyond the libidinal zone and includes developments in the ego as it confronts the social world.

2. The Anal Stage

Zones and modes. At Freud's second stage, between the ages of about one and a half and three years, the anal zone comes into prominence. With the maturation of the nervous system, children gain voluntary control over their sphincter muscles; they can now retain and eliminate as they wish. They often hold on to their bowels to maximize the sensations of the final release.

Erikson agrees with Freud that the basic modes of this stage are retention and elimination, of holding on and letting go. However, Erikson also points out that these modes encompass more than the anal zone. For example, children begin to use their hands to hold stubbornly onto objects and, just as defiantly, to throw them away. Once they can sit up securely, they carefully pile things up one moment, only to discard them the next. With people, too, they sometimes hold on, snuggling up, and at other times insist on pushing the adult away (Erikson, 1959, p. 82, 86).

The general stage: Autonomy vs. Shame and Doubt. Amidst these contradictory impulses—holding on one moment and expelling the next—the child is primarily trying to exercise a choice. Two-year-olds want to hold on when they want and to push aside when they do not. They are exercising their will, their sense of autonomy.

In other ways, too, maturation ushers in a sense of autonomy during the second and third years. Children can now stand up on their own two feet, and they begin to explore the world on their own. They also insist on feeding themselves, even if this means exercising their right to make a mess. Their language, too, reveals a new-found autonomy and sense of selfhood; they repeatedly use the words "me" and "mine." Most of all, they express their autonomy in a single word—"no." Two-year-olds seem unable to say "yes," as if any agreement means a complete forfeiture of their independence. Through the strong and insistent "no," children defy all external control.

As children seem so much more in control of themselves and reach peaks of willfulness, societies, through parents, decide it is time to teach them the right ways to behave. As Freud observed, parents do not permit their children to enjoy their anality in any way they please; instead they train them to behave in the socially proper way. Parents quite often toilet-train children by making them feel ashamed of messy and improper anal behavior. Children may resist training for some time, but they eventually submit to it.

Erikson agrees that the "battles of the toilet-bowel" are important. But he also is suggesting that the battle of wills—between the child's autonomy and the society's regulations—takes place in a number of arenas. For example, when children insist on feeding themselves and making a mess, parents try to regulate their behavior. Similarly, parents sooner or later decide that their two-year-olds cannot say "no" to every single request. Two-year-olds, like everyone else, must live in society and respect others' wishes. Thus, the conflict at this stage is a very general one.

Erikson defines the conflict as that of autonomy versus shame and doubt. Autonomy comes from within; biological maturation fosters the ability to do things on one's own—to control one's own sphincter muscles, to stand on one's own feet, to use one's hands, and so on. Shame and doubt, on the other hand, come from an awareness of social expectations and pressures. Shame is the feeling that one does not look good in others' eyes. For example, a little girl who wets her pants becomes self-conscious, worried that others will see her in this state. Doubt stems from the realization that one is not so powerful after all, that others can control one and perform actions much better.

Hopefully, children can learn to adjust to social regulations without losing too much of their initial sense of autonomy. Parents in some cultures try to help the child with this. They try gently to help the child learn social behavior without crushing the child's will. Other parents, unfortunately, are not so sensitive. They may shame children excessively when they have a bowel accident; they may try to break their children of any oppositional behavior; or they may ridicule their children's efforts to do things on their own. In such instances, children can develop lasting feelings of shame and doubt which override their impulses toward self-determination.

3. The Phallic (Oedipal) Stage

Zone and modes. During Freud's third stage (between about three and six years), the child's concern with the anal zone gives way to the primacy of the genital zone. Children now focus their interest on their genitals and become curious about the sex organs of others. They also begin to imagine themselves in adult roles and even dare to rival one parent for the love of the other. They enter the oedipal crisis.

Erikson calls the primary mode at this stage *intrusion.* By this term, he hopes to capture Freud's sense of the child as now exceedingly daring, curious, and competitive. The term intrusion describes the activity of the boy's penis, but as a general mode it refers to much more. For both sexes, the maturation of physical and mental abilities impels the child forward into a variety of intrusive activities. "These include the intrusion into other bodies by physical attack; the intrusion into other people's ears and minds by aggressive talking; the intrusion into space by vigorous locomotion; the intrusion into the unknown by consuming curiosity" (1950, p. 87).

The general stage: Initiative vs. Guilt. Initiative, like intrusion, connotes forward movement. The child with a sense of initiative makes plans, sets goals, and perseveres in attaining them. I noted, for example, some of the activities of one of our sons when he was five years. In a single day, he decided to see how high he could build his blocks, invented a game which consisted of seeing who could jump the highest on his parents' bed, and led the family to a new movie containing a great deal of action and violence. His behavior had taken on a goal-directed, competitive, and imaginative quality.

The crisis comes when children realize that their biggest plans and fondest hopes are doomed for failure. These ambitions, of course, are the oedipal ones— the wish to possess one parent and rival the other. The child finds out that these wishes violate deep social taboos and that they are far more dangerous than imagined. Consequently, the child internalizes social prohibitions—a guilt-producing superego—to keep such dangerous impulses and fantasies in check. The result is a new form of self-restriction. Forever after, the individual's naive exuberance and daring will be offset by self-observation, self-control, and self-punishment.

In Erikson's view, the creation of a superego constitutes one of the great tragedies in life. Although the superego is necessary for socialized behavior, it stifles the bold initiative with which the child met life at the phallic stage. Still, Erikson is not completely pessimistic. He observes that three- to six-year-old children are, more than at any other time, ready to learn quickly and avidly, and they are willing to find ways of channeling their ambition into socially useful pursuits (1950, p. 258). Parents can help this process by easing their authority somewhat and by permitting children to participate with them as equals on interesting projects. In this way, children do not have to give up their ambition altogether and can begin to attach it to the goals of adult social life.

4. The Latency Stage

In Freud's theory, the resolution of the Oedipus complex brings about a latency period, lasting from about six to 11 years. During this period, the sexual and aggressive drives, which produced crises at earlier periods, are temporarily dormant. There is no libidinal (sexual) zone for this stage.

Of course the child's life at this time may not be entirely conflict-free. For example, the birth of a sibling may arouse intense jealousy. But as a rule, this is a period of calm and stability. In terms of the instincts and drives, nothing much is going on.

Erikson, however, shows that this is a most decisive stage for ego growth.

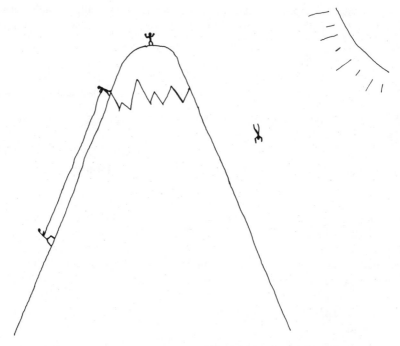

FIGURE 8.1 A boy's drawing expresses daring initiative and disaster. There is a resemblance to the myth of Icarus, about a boy who went too far—too near the sun. This drawing, by a nine-year-old, may also suggest the industry theme of the next stage; the climbers on the left are engaged in skillful cooperation.

Children master important cognitive and social skills. The crisis is *industry versus inferiority*. Children forget their past hopes and wishes, which were often played out within the family, and eagerly apply themselves to learning the useful skills and tools of the wider culture. In preliterate societies, children learn to hunt, farm, and make utensils. In these cultures, learning is often informal, and much of it comes from older children. In modern technological societies such as ours, children go to school, where they are asked to master more cerebral skills— reading, writing, and arithmetic. In either case, they are learning to do meaningful work and are developing the ego strengths of "steady attention and persevering diligence" (1950, p. 259). They also are learning to work and play with their peers.

The danger of this stage is an excessive feeling of inadequacy and inferiority (1950, p. 260). Most of us can probably remember the hurts of failure in the classroom or on the playground. A particularly deep sense of inferiority may have various roots. Sometimes children have difficulty at this stage because they have not successfully resolved the conflicts at earlier stages. For example, a girl may have developed more doubt than autonomy at the second stage, so she is unsure of herself as she tries to master new tasks. At other times, school and community attitudes may hinder the development of a sense of industry. A black boy, for example, may learn that the color of his skin counts for more than his wish and will to learn. And it is all too often that schools fail to discover and encourage the individual's special talents (1950, p. 260).

On the other hand, good teachers (who often are those who feel trusted and respected by the community) can help children at this time. "Again and again," Erikson says, "I have observed in the lives of especially gifted and inspired people that one teacher, somewhere, was able to kindle the flame of a hidden talent" (1959, p. 87).

5. Puberty (The Genital Stage)

According to Sigmund and Anna Freud, adolescence is a turbulent stage because of the dramatic physiological changes occurring at this time. Sexual and aggressive drives, which were dormant during the latency stage, now threaten to overwhelm the ego and its defenses. The genital zone, in particular, is infused with tremendous sexual energy, and the adolescent is once again troubled by oedipal fantasies. The teenager may find it difficult simply to be around his or her parents.

Erikson agrees that the great increase in drive energy is disruptive at adolescence, but he sees this as only part of the problem. Adolescents also become disturbed and confused by new *social* conflicts and demands. The adolescent's primary task, Erikson thinks, is establishing a new sense of *ego identity*—a feeling for who one is and one's place in the larger social order. The crisis is one of *identity versus role confusion.*

The upsurge of instinctual drives certainly contributes to the adolescent's identity problems. Adolescents suddenly feel as if their impulses have a will of their own, that they are no longer one with themselves. Also, the rapid physical growth at puberty creates a sense of identity confusion. Young people begin to grow so quickly and to change in so many ways that they barely recognize themselves. It is probably for this reason that teenagers spend so much time looking in mirrors and pay so much attention to their appearance.

But identity problems are as much, if not more, a social matter. It is not physical growth or sexual impulses *per se* which trouble young people, but it is the thought that one might not look good to others or meet others' expectations. And, even more than this, young people begin to worry about their future place in the larger social world. Adolescents, with their rapidly expanding mental powers, feel overwhelmed by the countless options and alternatives before them.

Because adolescents are so uncertain about who they are, they anxiously tend to identify with "in groups." They can "become remarkably clannish, intolerant, and cruel in their exclusion of others who are 'different' . . ." (1959, p. 92). In their hurry to find some identity, they stereotype "themselves, their ideals, and their enemies" (p. 92). Some youth align themselves with ideologies— national, political, or religious—which provide them with a group identity and clear-cut images of good and bad in the world.

Actually, identity-formation is a life-long process. In part, we form our identities through identifications. Although we are not necessarily aware of it, we identify with those who appeal to us and therefore become like them. Each person's identity, then, is partly a synthesis of various partial identifications (1959, pp. 112-13).

We also develop a sense of identity through our accomplishments. The ability to stand up, walk, run, play ball, draw, read, and write all contribute to a sense of ego identity. We come to see ourselves as "one who can do these things." Such accomplishments become part of a positive and lasting sense of identity when they have importance for the culture (1959, pp. 89-90).

Still, even though identity-formation is a life-long process, the problem of identity reaches its crisis at adolescence. It is at this time that so many inner changes are taking place, and so much in terms of future commitment is at stake. At this time, one's earlier identity seems inadequate for all the choices and decisions one must make.

Erikson (1959, p. 123) asks us to consider, for example, a young college girl from a conservative background. When she goes to college she meets people of very different backgrounds, among whom she must choose her friends. She also must decide what her attitude toward sex will be and what occupational goals she is willing to compete for. At this point, her previous identity and identifications offer little help. Each decision seems to affirm some aspect of her past while repudiating others. If, for example, she decides to become sexually active, she may violate her family's spoken values, while identifying with some of their hidden wishes. Similarly, if she chooses to compete in a male-dominated profession, such as medicine or law, she ventures beyond certain family values but aligns herself with others. Thus, as she makes decisions and commitments, she reworks prior identifications and forms a new identity. Her task is to forge for herself some "central perspective and direction, some working unity, out of the effective remnants of [her] childhood and the hopes of [her] anticipated adulthood" (Erikson, 1958, p. 14).

Identity formation is largely an unconscious process. Still, young people are often painfully aware of their inability to make lasting commitments. They feel that there is too much to decide too soon, and that every decision reduces their future alternatives (Erikson, 1959, p. 124). Because commitment is so difficult, they sometimes enter a *psychosocial moratorium,* a kind of "time out" period for finding oneself. Some young people, for example, drop out of college to travel or experiment with various jobs before making any final decisions. However, many young people have trouble achieving a free moratorium state. Until they know who they are and what they will do in life, they often experience a sense of isolation, a feeling that time is passing them by, an inability to find meaning in any kind of activity, and a feeling that life is simply happening to them (1959, p. 126). Nevertheless, the adolescent frequently postpones commitments anyway, because of an inner need to avoid *identity foreclosure,* a premature acceptance of compartmentalized social roles. And although a protracted identity search can be painful, it can also eventually lead to a higher form of personal integration and to genuine social innovations. As we have seen, many of the theorists in this book resisted the temptation to settle into conventional occupational identities. Piaget, Freud, and Erikson, among others, spent some time searching for their true callings. And their search, while not always pleasant for them, eventually led to new ways of understanding and meaningful changes in their professions.

TABLE 8.1   The Stages of Freud and Erikson

Age	*Freud's Stage*	*Erikson's General Stage*
Birth to one and a half	Oral	Trust vs. Mistrust
One and a half to three	Anal	Autonomy vs. Shame, Doubt
Three to six	Phallic (Oedipal)	Initiative vs. Guilt
Six to 11	Latency	Industry vs. Inferiority
Adolescence	Genital	Identity vs. Role Confusion
Young Adulthood		Intimacy vs. Isolation
Adulthood		Generativity vs. Stagnation
Old Age		Ego Integrity vs. Despair

6. Young Adulthood

Erikson is the first Freudian and one of the few developmental writers of any persuasion to propose separate stages for the adult years. If, then, his thoughts seem sketchy, we should remember that he is writing about an uncharted area.

Erikson's stages of adult development describe steps by which people widen and deepen their capacities to love and care for others. The *adolescent* is preeminently self-absorbed. Adolescents are concerned with who they are, how they appear in the eyes of others, and what they will become. They do become sexually attracted to others and even fall in love, but such attachments most often are really efforts at self-definition. In their interactions, young people try to find out who they are by endlessly talking about their true feelings, their views of each other, and their plans, hopes, and expectations (Erikson, 1959, p. 95).

The adolescent, then, is too preoccupied with who he or she is to take up the task of young adulthood—the attainment of *intimacy*. Real intimacy is only possible once a reasonable sense of identity has been established (1959, p. 95). Only one who is secure with one's identity is able to lose oneself in true mutuality with another. The young man, for example, who is worried about his masculinity will not make the best lover. He will be too self-conscious, too worried about how he is performing, to abandon himself freely and tenderly with his sexual partner. To the extent that people fail to attain genuine mutuality, they experience the opposite pole of this stage—*isolation and self-absorption.*

Erikson observes that some young people marry before they have established a good sense of identity. The hope that they will be able to find themselves in their marriage. Such marriages, however, rarely work out. The partners sooner or later begin to feel hemmed in by their obligations as mates and parents. They soon complain that the others are not giving them the opportunity to develop themselves. Erikson notes that a change in mate is rarely the answer. What the individual needs is some "wisely guided insight" into the fact that one cannot expect to live intimately with another until one has become oneself (1959, p. 95).

In his discussions of intimacy, Erikson, as a respectable Freudian, speaks glowingly of the experience of orgasm. It is a supreme experience of mutual regulation which takes the edge off the inevitable bitternesses and differences between two people (1950, p. 265). He adds, though, that the "Utopia of genitality" is by no means a purely sexual matter. True intimacy means that two people are willing to share and mutually regulate all important aspects of their lives (1950, p. 266).

7. Adulthood

Once two people have established some measure of intimacy, their interests begin to expand beyond just the two of them. They become concerned with raising the next generation. In Erikson's terms, they enter the stage of *generativity versus stagnation.* "Generativity" is a broad term, which refers not only to the creation and care for children, but the production of things and ideas through work. However, Erikson has focused more on the former—the care for children.

The mere fact of having children, of course, does not guarantee generativity. Parents must do more than produce offspring; they must adequately care for and guide them. At the same time, people can achieve a sense of generativity without having children of their own. Nuns and priests, for example, forego the right to raise their own children, as do others who apply their special gifts to other areas. Such persons can still guide and care for the next generation "by working with other people's children or helping to create a better world for them" (Erikson in Evans, 1969, p. 51). Such adults must, to be sure, withstand a certain amount of frustration. This is especially true of women, whose bodies are built for nurturing and nourishing offspring. However, a sense of generativity is still possible.

When generativity is lacking, the result is stagnation and impoverishment of the personality. In such cases, people often regress to a kind of "pseudo-intimacy" or "begin to indulge themselves as if they were their one and only child" (Erikson, 1959, p. 97). Erikson might be thinking of couples who endlessly analyze their relationships in terms of how much each person is getting from the other. They seem more concerned with their own needs than those of their children.

There are several possible reasons for an inability to develop at this stage. Sometimes the parent's own childhood was so empty or frustrating that the parent cannot see how it is possible to do more for his or her children. In other cases, the difficulty seems more cultural. In the United States in particular, our values emphasize independent achievement to such an extent that people can become too exclusively involved in themselves and their successes and neglect the responsibility of caring for others (1959, p. 97).

8. Old Age

The psychological literature on old age, which is still sparse, typically views this period as one of decline. The elderly, it is repeatedly pointed out, must cope with a series of physical and social losses. They lose their physical strength and health, they lose their jobs and much of their income through retirement, and, as time goes by, they lose their spouses, relatives, and friends. Equally damaging, they suffer the inevitable loss of status that accompanies being old, inactive, and "useless" in America. Occasionally, authors acknowledge that the aged are supposed to gain an inner wisdom, but this is probably only the case in some far-off culture or in some time long past. By and large, successful aging consists in somehow adjusting to physical and social setbacks (see Sze, 1975, pp. 569-74; Havighurst, 1952 and 1968; Gitelson, 1975).

Erikson is aware of the many adjustments, physical and social, which the elderly must make. He is aware of the fact that the aged cannot be as active as they once were. But his emphasis is different. It is not on external adjustments but on the inner struggle of this period—a struggle which holds potential for growth and even wisdom. He calls this struggle *ego integrity versus despair.*

As older people face death, they engage, Erikson implies, in what has been called a life review (Butler, 1963). They look back on their lives and wonder whether they were worthwhile. In this process, they confront the ultimate despair—the feeling that life was not what it should have been, but now time has run out and there is no chance to try alternative life styles. Frequently, disgust hides despair. Many older people are disgusted by every little thing; they have no patience for the struggles and failings of others. Such disgust, Erikson says, really signifies their contempt for themselves (1959, p. 98).

As the older person faces despair, he or she is trying to find a sense of ego integrity. Ego integrity, Erikson says, is difficult to define but includes "the acceptance of one's one and only life cycle as something that had to be and that, by necessity, permitted of no substitutions . . ." (1950, p. 268). Integrity, it would seem, expresses the feeling that, "Yes, I made mistakes, but given who I was at the time and the circumstances, the mistakes were inevitable. I accept them, along with the good things in my life." Integrity is a feeling which also extends beyond the self and even transcends national and ideological boundaries. The older person, on some level, has a feeling of companionship "with the ordering ways of distant times and different pursuits, so expressed in the simple products and sayings of such times and pursuits" (1950, p. 268). Ego integrity, then, leads to a kind of detached and philosophical wisdom about life itself.

Erikson tells us (1976) that the crisis of old age is most admirably illustrated by Ingmar Bergman's film, *Wild Strawberries.* The film, in Erikson's words,

> records an old Swedish doctor's journey by car from his place of retirement to the city of Lund. There, in the ancient cathedral, Dr. Isak Borg is to receive the highest honor of his profession, a Jubilee Doctorate marking fifty years of meritorious service. But this journey by car on marked roads through familiar territory also becomes a symbolic pilgrimage back into childhood and deep into his unknown self (1976, p. 1).

The film begins with Borg writing in his diary and then plunges into a terrifying dream symbolizing his fear of death. Upon awakening, Borg decides to travel to Lund by car instead of by airplane and to take along his daughter-in-law, Marianne, who is in the midst of a marital crisis with which Borg has so far refused to help. As soon as they are in the car, they begin to quarrel, and Marianne tells him that "even though everyone depicts you as a great humanitarian . . . you are an old egotist, Father" (Bergman, 1957, p. 32). Along the journey, Borg engages in other encounters with Marianne and others, and he is visited by vivid dreams and memories about the past. These dreams and memories are extremely humiliating to him. He comes to realize that throughout his life he has been an isolated onlooker, moralistically aloof, and in many ways incapable of love. We see, then, that Borg's initial sense of integrity was superficial; as he imagines death and reviews his life, he confronts its many failures.

In the end, however, Borg's insights do not lead to a final despair but to a new kind of wisdom and an acceptance of the past. While he is receiving his Jubilee Doctorate, which by now has become a rather trivial event, he begins to see "a remarkable causality" in the events of his life—an insight which sounds remarkably similar to Erikson's statement that ego integrity includes a sense of the inevitable order of the past. Even more impressive, though, is a change in character. At the end of the film, Borg expresses his love for Marianne and offers to help her and his son.

Through this film, we see why Erikson emphasizes the importance of both the positive and negative poles of his crises. Borg's initial sense of integrity was superficial and unconvincing. He acquired a meaningful sense of integrity only after confronting his life thoroughly and answering to some existential despair (Erikson, 1976, p. 23).

Erikson and Bergman, then, are pointing to an inner struggle which we are apt to miss when we look at older people. We are aware of their many physical and social difficulties, and we may deplore the fact that older people seem so "useless." But such judgments are only partly valid. They are opinions formed by looking only at external behavior. We see that older people lack the zest and youthfulness which we so greatly prize, but we fail to consider the *inner* struggle. We fail to see that the quiet older person may be grappling in some way with the most important of all questions: Was my life, as I face death, a meaningful one? What makes a life meaningful?

THEORETICAL ISSUES

Why Erikson's Theory Is a Stage Theory

Erikson's theory, like Piaget's, meets the formal requirements of a stage theory. That is, Erikson's stages 1) describe general issues at each period, 2) refer to qualitatively different conflicts, 3) unfold in an invariant sequence, and 4) are culturally universal. Let us examine each of these points in turn.

1. The stages describe general issues. As we have emphasized, stages refer to general characteristics or issues. Accordingly, Erikson has tried to go beyond Freud's relatively specific focus on body zones and has attempted to delineate the general issues at each period. For example, at the oral stage, he shows that it is not just the stimulation of this zone that is important but the general mode of taking in, and, more generally still, the development of a sense of trust in one's providers. Similarly, at each stage Erikson tries to pinpoint the most general issue faced by the individual in the social world.

2. The stages refer to qualitatively different behavior patterns. If development were just a matter of gradual quantitative change, any division into stages would be arbitrary. Erikson's stages, however, give us a good sense of how behavior is qualitatively different at different points. Children at the autonomy stage sound very different from those at the trust stage; they are much more independent. Children at the initiative stage are different again. Whereas children who are establishing a sense of autonomy defy authority and keep others out, children with a sense of initiative are more daring and imaginative, running vigorously about, making big plans, and initiating new activities. Similarly, behavior has a distinctive flavor at each stage.

3. The stages unfold in an invariant sequence. All stage theories imply an invariant sequence, and Erikson's is no exception. He says that each stage is present in some form throughout life but that each reaches its own crisis at a specific time and in a specific order.

Erikson's claim is based on the assumption that his sequence is partly the product of *biological maturation.* As he puts it, the child obeys "inner laws of development, namely those laws which in his prenatal period had formed one organ after another and which now create a succession of potentialities for significant interaction with those around him" (1950, p. 67). At the second stage, for example, biological maturation ushers in a sense of autonomy. Because of maturation, the child can stand on his or her own two feet, control his or her sphincter muscles, walk, use words such as "me," "mine," and "no," and so on. At the third stage, maturation prompts a new sexual interest, along with capacities for imaginative play, curiosity, and vigorous locomotion.

At the same time, *societies* have evolved such that they invite and meet this inner, maturational succession of potentialities. When, for example, the child at the autonomy stage demonstrates a new degree of self-control, socializing agents consider the child ready for training. For example, they begin toilet-training. The result is the battle of wills, between child and society, which creates the crisis of this period. Similarly, when children become recklessly ambitious with respect to sexual matters, societies decide it is now time to introduce their particular sexual prohibitions, creating the core conflict at the third stage. Thus, the succession of crises is produced by inner maturation on the one hand and social forces on the other.

4. The stages are cultural universals. Erikson believes that his stages can be applied to all cultures. Readers are frequently skeptical of this claim, for they know how widely cultures differ.

Erikson, too, is aware of the vast differences among cultures. In fact, one of his goals is to show how cultures handle the stages differently according to their different value-systems. For example, the Sioux provide their children with a long and indulgent period of nursing; one of their goals is to get children to trust others and to become generous themselves (Erikson, 1950, pp. 134-40). Our society, in contrast, discourages dependency. Compared to other cultures, we wean our infants very early. We do not seem to want our children to learn to depend on or trust others too much, but to become independent. Independence and free mobility seem part of our cultural ethos, from the pioneer days to the present time (1950, Ch. 8).

What Erikson does claim is that all cultures address themselves to the *same issues.* All cultures try to provide their children with consistent care, regulate their extreme wish to do everything their own way, and instill incest taboos. And, as children grow, all cultures ask them to learn the tools and skills of their technology, to find a workable adult identity, to establish bonds of intimacy, to care for the next generation, and to face death with integrity. All cultures attempt to achieve these tasks because culture itself is a part of the evolutionary process; in the course of evolution, those groups which failed in these tasks had less chance of surviving. For example, unless cultures could get their members to sacrifice some of their independence for the needs of others (at the autonomy stage), to begin to learn the skills and tools of the society (at the industry stage), and to care for the next generation (at the generativity stage), they probably did not endure.[1]

Must One Go Through All the Stages?

We sometimes hear that if one doesn't achieve a good measure of success at one of Erikson's stages, one may be unable to go on to the next stage. This is wrong. In Erikson's theory, one must, if one lives long enough, go through all the stages. The reason has to do with the forces that move one from stage to stage: biological maturation and social expectations. These forces push one along according to a certain timetable, whether one has been successful at earlier stages or not.

Consider, for example, a boy who has been unable to attain much of a sense of industry. When he reaches puberty, he must grapple with the issues of identity even though he is not really ready to do so. Because of biological changes, he finds himself troubled by an upsurge of sexual feelings and by a rapidly changing body. At the same time, social pressures force him to cope with problems of dating and to start thinking about his future occupation. It matters little to the larger society that he is still unsure about his own skills. His society has its own timetable, and by the time he is 20 years or so, he will feel pressure to decide on a career. In the same way, he will find himself confronting each new stage in the sequence.

Each person, then, must go through all the stages, whether he or she has traversed the earlier stages well or not. What is true is that success at earlier stages affects the chances of success at later ones. This is why the first stage, trust versus mistrust, is so important. Children who developed a firm sense of trust in their caretakers can afford to leave them and independently explore the environment. Their strong autonomy, in turn, naturally leads toward a vigorous initiative, and so on. Children, in contrast, who lack trust—who are afraid to let caretakers out of sight—are less able to develop a sense of autonomy and succeeding strengths. They also will be more prone to regress to earlier stages when difficulties arise. Nevertheless, developmental forces will require them to face the issues of each new stage.

The situation is different, then, from that in Piaget's sequence, where a person who has little curiosity about an area (e.g., mathematics) may never

---

[1] See the discussion of Darwin, Ch. 3.

reach the highest stages of thinking in that area. In Erikson's theory, biological maturation and social pressures force the individual to face the crisis at each succeeding stage in the sequence.

Comparison with Piaget

We have now reviewed the two most influential stage theories in the developmental literature: Piaget's cognitive-developmental theory and Erikson's psychoanalytic theory. In many respects, the two theories seem very different. As we have just mentioned, they have different ideas about the forces behind developmental change. Even more noticeably, Piaget and Erikson seem to be writing about different areas of experience: Piaget focuses on intellectual development and the child's understanding of the physical world; Erikson concentrates on emotional development and the child's interactions with people.

Yet, despite these and other differences, it also seems that Piaget and Erikson are often really presenting different perspectives on the same basic developments at each major phase of childhood. The following notes, while incomplete, will hopefully suggest how this is so.

1. Trust. As Erikson observes (1964, pp. 116-17), both Piaget and he are concerned with the infant's development of a secure image of external objects. Erikson points to the child's growing reliance on the predictability and dependability of people, whereas Piaget documents the developing sense of permanent things. Thus, both are concerned with the child's growing faith in the stability of the world. Several researchers have begun exploring these parallels (Bell, 1970; Gouin-Décarie, 1965; Flavell, 1977, p. 54).

2. Autonomy. As children develop a sense of trust in their caretakers, they become increasingly independent. Secure in their knowledge that others will be there when needed, they are free to explore the world on their own.

Piaget points to a similar process. As children gain the conviction that objects are permanent, they can act increasingly independently of them. For example, when his daughter Jacqueline's ball rolled under the sofa, she was no longer bound to the spot where she last saw it. She now knew that the object was permanent, even if hidden, and could therefore try out alternative routes for finding it.[2]

3. Initiative. At this stage, between about three and six years, Erikson and the Freudians emphasize the child's consuming curiosity, wealth of fantasy, and daring imagination. As Erikson says, "Both language and locomotion permit

[2] For further thoughts on the parallels at this stage, see Kohlberg and Gilligan, 1971, p. 1076.

him to expand his imagination over so many things that he cannot avoid frightening himself with what he has dreamed and thought up" (1959, p. 75).

Piaget's view of the thinking at this period is remarkably similar. As Flavell says,

> The preoperational child is the child of wonder; his cognition appears to us naive, impression-bound, and poorly organized. There is an essential lawlessness about his world without, of course, this fact in any way entering his awareness to inhibit the zest and flights of fancy with which he approaches new situations. Anything is possible because nothing is subject to lawful constraints (1963, p. 211).

For Piaget, then, the fantasy and imagination of the phallic-age child owes much to the fact that the child is in the preoperational period—a time in which thoughts run free because they are not yet tied to the systematic logic which the child will develop at the next stage.

4. Industry. For Erikson and the Freudians, the fantasies and fears of the oedipal child are temporarily buried during the latency stage, which lasts from about seven to 11 years. Frightening wishes and fantasies are repressed, and the child's interests expand outward; the child intently tries to master the realistic skills and tools of the culture. In general, this is a relatively calm period; children seem more self-composed.

Piaget, too, would lead us to believe that the six- to 11-year-old is more stable, realistic, and organized than the younger child. For Piaget, this change is not the result of the repression of emotions and dangerous wishes; rather, it comes about because, intellectually, the child has entered the stage of concrete operations. The child can now separate fact from fancy, can see different perspectives on a problem, and can work logically and systematically on concrete tasks. Intellectually, then, the child is in a stage of equilibrium with the world, and this contributes to his or her overall stability and composure. Erikson himself seems to have concrete operations in mind when he describes this period: He says that at this time the child's "exuberant imagination is tamed and harnessed by the laws of impersonal things . . ." (1950, p. 258).

5. Identity. In Erikson's view, the calm of the preceding period gives way to the turbulence and uncertainty of adolescence. Adolescents are confused by physical changes and pressures to make social commitments. They wonder who they are and what their place in society will be.

Piaget has little to say about physical changes in adolescence, but his insights into cognitive development help us understand why this can be an identity-searching time. During the stage of concrete operations, the child's thought was pretty much tied to the here-and-now. But with the growth of formal operations, the adolescent's thought soars into the distant future and into the realm of the purely hypothetical. Consequently, adolescents can now entertain limitless possibilities about who they are and what they will become. Formal operational capacities, then, enable them to raise questions about their identity (see Inhelder and Piaget, 1955, Ch. 18).

PRACTICAL IMPLICATIONS

Clinical Work: A Case Illustration

Clinical psychologists and other mental health workers have found Erikson's concepts very useful. We can get a sense of this from Erikson's own work with one of his cases, a four-year-old boy he calls Peter.

Peter suffered from a psychogenic megacolon, an enlarged colon which resulted from Peter's emotionally based habit of retaining his fecal matter for up to a week at a time. Through conversations with Peter and his family, Erikson learned that Peter developed this symptom shortly after his nurse, an Asian girl, had been dismissed. Peter, it seems, had begun "attacking the nurse in a rough-housing way, and the girl had seemed to accept and quietly enjoy his decidedly 'male' approach" (Erikson, 1950, p. 56). In her culture, such behavior was considered normal. However, Peter's mother, living in our culture, felt there was something wrong about Peter's sudden maleness and the way the nurse indulged it. So she got rid of the nurse. By way of explanation, the nurse told Peter that she was going to have a baby of her own, and that she preferred to care for babies, not big boys like Peter. Soon afterward, Peter developed the megacolon.

Erikson learned that Peter imagined that he himself was pregnant, a fantasy through which he tried to keep the nurse by identifying with her. But, more generally, we can see how Peter's behavior regressed in terms of stages. Initially, he had begun displaying the attacking, sexual behavior of the initiative stage, but he found that it led to a tragic loss. So he regressed to an anal modality. He was expressing, through his body, his central need: *to hold on.* When Erikson found the right moment, he interpreted Peter's wishes to him, and Peter's symptom was greatly alleviated.

Sometimes students, upon hearing of Peter's behavior, suggest that his symptom was a means of "getting attention." This interpretation is used frequently by the behaviorists. We note, however, that Erikson's approach was different. He was concerned with the meaning of the symptom for Peter, with what Peter was trying to express through it. Through his body, Peter was unconsciously trying to say, "I need to hold on to what I've lost." Erikson and other psychoanalysts believe that instead of changing a child's behavior through external reinforcements such as attention, it is best to speak to the child's fears and to what the child may be unconsciously trying to say.

Thoughts on Child-Rearing

Over the years, Erikson has applied clinical insights to many problems, including those in education, ethics, and politics. He also has had a special interest in child-rearing.

As we briefly mentioned in our discussion of trust, Erikson is concerned with the problem facing parents in our changing society. American parents often are unable or unwilling simply to follow traditional child-rearing precepts; they would like to bring up their children in more personal, tolerant ways, based on new information and education (Erikson, 1959, p. 99). Unfortunately, modern child-rearing advice is often contradictory and frightens the new parent with its accounts of how things can go wrong. Consequently, the new parent is anxious and uncertain. This is a serious problem, Erikson thinks, for, as we have seen, it is important that the parent convey to the child a basic security, a feeling that the world is a calm and secure place.

Erikson suggests that parents can derive some inner security from religious faith. Beyond this, he suggests that parents heed their fundamental "belief in the species" (1950, p. 267). By this, Erikson means something similar to Gesell. Parents should recognize that it is not all up to them to form the child; children largely grow according to an inner, maturational timetable. As Erikson says, "It is important to realize that . . . the healthy child, if halfway properly guided, merely obeys and on the whole can be trusted to obey inner laws of development . . ." (1950, p. 67). Thus, it is all right for parents to follow their inclination to smile when their baby smiles, make room for their child to walk when he or she tries to, and so on. They can feel secure that it is all right to follow the baby's own biological ground plan.

Erikson also hopes that parents can recognize the basic inequality between child and adult. The human child, in contrast to the young of other species, undergoes a much longer period of dependency and helplessness. Parents, therefore, must be careful to resist the temptation to take out their own frustrations on the weaker child. They must resist, for example, the impulse to dominate the child because they themselves feel helpless with others. Parents also should be careful to avoid trying to shape the child into the person they wanted to become, thereby ignoring the child's own capacities and inclinations. Erikson says, in conclusion, "If we will only learn to let live, the plan for growth is all there" (1959, p. 100).

EVALUATION

Erikson has certainly broadened psychoanalytic theory. He has delineated the most general issues at each of Freud's stages and has enlarged the stage sequence so that it now covers the entire life cycle. Erikson also has given us a new appreciation of how social factors enter into the various stages. For example, he shows that adolescents are not just struggling to master their impulses, but are trying to find an identity in the larger social world.

Erikson, finally, has given Freudian theory new insights into the possibilities for healthy development. He has primarily done this by making wider use of the concept of maturation than Freud did. In Freud's view, maturation directs the course of the instinctual drives, which must undergo a good measure of repression. For Erikson, maturation also promotes the growth of the ego modes and the general ego qualities such as autonomy and initiative.[3] Erikson, to be sure, discusses the difficulties in attaining these qualities, but he does give us a better picture of how ego growth is possible. By suggesting that healthy development is tied to a maturational ground plan, Erikson has moved Freudian theory in the developmental direction of Rousseau, Gesell, and others.

Erikson's theory also has met with various criticisms. Robert White (1960) argues that Erikson has tried too hard to link the various aspects of ego development to Freud's libidinal zones. Erikson says that for each zone, there is a characteristic ego mode of interaction with the world. However, White argues, these modes fail to capture many of the child's activities. For example, many of the young child's efforts to achieve autonomy—such as the child's loud "no's" and vigorous walking—seem unrelated to the anal modes of retention and elimination. White himself proposes that we think of ego growth as a general tendency toward competence—a tendency which includes locomotion, exploration, and autonomous action without any necessary connection to Freud's zones.

A more general criticism is that Erikson is frequently vague with respect to theoretical issues. He writes in a beautiful flowing prose, but he leaves many conceptual matters unclear. For example, he provides new insights into the potential for growth in old age, when people examine their lives and search for wisdom, but he does not clearly indicate how this is part of the maturational process. It may be that there is a biologically based tendency to review one's life (Butler, 1963), but Erikson is not explicit on this matter. Similarly, he fails to spell out how maturation contributes to the other stages of adulthood, and he treats many other issues in a very implicit way.

Erikson is aware of his general vagueness. As he once said, "I came to psychology from art, which may explain, if not justify, the fact that the reader will find me painting contexts and backgrounds where he would rather have me point to facts and concepts" (1950, p. 17).

This conceptual vagueness may partly explain why interesting empirical research on his theory has been slow to develop. Nevertheless, some good research is beginning to emerge. Marcia (1966), for example, has thoughtfully constructed measures of different identity states, and these measures seem related to other

---

[3] Erikson's suggestion that ego growth has maturational roots follows the lead of Hartmann, discussed in the preceding chapter (pp. 140-41).

variables. For example, young people with a foreclosed identity—who have simply accepted handed-down occupational goals and values without themselves struggling with alternatives—seem to be most often found at the level of conventional moral thought on Kohlberg's scale. Those, in contrast, who have achieved a sense of identity after a personal struggle are more often represented by postconventional moral thinking (Podd, 1972).

Thus, it does appear that Erikson's theory can generate good empirical research. Erikson's thought, like Freud's, is so rich and profound that it rewards efforts to master it—both for personal insights into human nature and for scientific progress.

9

A Case Study
in Psychoanalytic Treatment:
Bettelheim on Autism

BIOGRAPHICAL INTRODUCTION

In this chapter, we will try to get a better understanding of psychoanalytic treatment. Such treatments vary, of course, and we will not attempt to review the variations here. Instead, we will examine a "case study": Bettelheim's work with autistic children. Bettelheim's work is especially important to us because, as we shall see, it is squarely within the developmental tradition.

Bruno Bettelheim was born in 1903 in Vienna, where he grew up and began his work as a psychoanalyst. Early on, he became interested in the disorder labeled "autism" (Kanner, 1943), a mysterious condition in which children are totally unresponsive to people. Before Bettelheim and his wife had children of their own, they cared for an autistic girl in their home. However, Hitler's invasion of Austria disrupted this treatment, along with everything else. From 1938 to 1939, Bettelheim was a prisoner in the concentration camps of Dachau and Buchenwald. Bettelheim has written extensive, brilliant accounts of this experience (e.g., 1960) and has drawn on it in all of his work. After his release, he came to the United States and, in 1944, took over the direction of the Orthogenic School in Chicago with the hope that if it is possible to build prison camps powerful enough to destroy human personalities, perhaps it also is possible to create environments that can foster their rebirth (Bettelheim,

1967, p. 8). In this special school and home for children, Bettelheim and his staff have treated a wide variety of emotional disorders, but Bettelheim has always had a special interest in autistic children. His accounts of his work with these children are primarily contained in his book, *The Empty Fortress* (1967).

THE AUTISTIC SYNDROME

Autism is relatively rare; it is much rarer, for example, than mental deficiency. However, it is important because it is the earliest of the severe personality disturbances, usually showing up by the second year of life (Treffert, 1970; Bettelheim, 1967, p. 392). Autistic children appear to be physically healthy, but they differ dramatically from normal children in other ways. As mentioned, these children are completely aloof; they do not interact with others. If you try to relate to them, they either avoid eye contact or "look right through you" (Lovaas, 1973).

Much of the time, these children engage in repetitive behavior called self-stimulation. That is, they endlessly spin objects, such as ashtrays, or stare at their hands as they rapidly shake them, or they rock themselves like babies. In a minority of cases, the children are self-destructive. For example, they hit their heads or tear at their skin, especially when placed in a new situation or when physically restrained (Lovaas *et al.*, 1973).

Autistic children, in addition, frequently display severe disturbances in language. In about half the cases, the children are mute. Those who do talk often engage in echolalia, the meaningless repetition of sounds. For example, if you ask a child, "What's your name?" the child answers, "What's your name?" (Lovaas, 1973, p. 2).

Another characteristic is an obsessive need for sameness. Once the child establishes a behavior pattern, the child insists on its ritualistic execution. For example, a bedtime ritual must always consist of precisely the same actions. If one interferes with the pattern, the child flies into a rage or into an acute panic (Eisenberg and Kanner, 1956).

Some of these symptoms appear in children with other psychiatric diagnoses, so one would not, for example, label a child "autistic" simply because he or she engaged in self-rocking. The most distinctive feature of autism is the extreme isolation.

THE CAUSE OF AUTISM

The cause of autism is still unknown. Because the onset is so early, many workers believe that it is a product of some inborn impairment, perhaps some brain dysfunction (e.g., Rimland, 1964). Others, including Bettelheim, speculate that autism is primarily the outcome of early interactions with the social environment, with parents or caretakers. More specifically, Bettelheim proposes that early in life these children fail to develop a sense of *autonomy*, which he defines as the feeling that one can have an effect on the environment.

According to Bettelheim, the issue of autonomy occurs much earlier in life than most writers have supposed. In the psychoanalytic tradition, infancy is frequently portrayed as a period of passive dependency, as a "golden age when all [the infant's] wants are taken care of by others and he neither wishes nor needs to do anything on his own" (Bettelheim, 1967, p. 15). Bettelheim counters that there are many ways in which babies ordinarily develop a sense of autonomy, a conviction that their own actions make a difference in the world. For example, babies are active in nursing, searching for the nipple on their own and expending great energy in sucking, and they probably feel that it is their own actions that produce the satisfactions of feeding. Similarly, babies' cries may give them a sense of autonomy and power. Although we adults may consider the babies' cries as a sign of their dependency, this probably is not the way babies see it. If their cries repeatedly bring comfort, they may get the sense that "When I cry, things happen." Also, babies may get a sense that their emotions and actions count for something when their smiles bring about smiles in return. Or, when babies are held too tightly, and their gestures of discomfort produce a more comfortable grip, they acquire further confirmation that their own actions make a difference. In Bettelheim's view, when we see children who act autonomously in the second year—for example, vigorously walking, exploring, and trying to do things on their own—they are only continuing to express a sense of autonomy that they began to develop in the first year of life (1967, pp. 14-32).

Bettelheim thinks that the baby's growing sense of autonomy not only is important in its own right but is central to the development of a personality and a self (1967, pp. 24-25, 33-35). For Bettelheim, the sense of self—the sense of "I"—is bound up with the feeling that "I did it," that "It was my own actions that made the difference."

Autism, Bettelheim thinks, is the failure to develop autonomy. Early on, autistic children seem to get the notion that their attempts to influence the key figures in their lives will meet only with indifference, anxiety, or retaliation. For example, they might conclude that their efforts to find the mother's nipple and to feed themselves produce mainly irritation or anxiety. Or, when they show discomfort at being held too tightly, they might experience the caretaker's reaction as one of indifference (or a further tightening of the hold). If such reactions dominate the infants' experience, they will begin to give up their own active efforts to better their lot (1967, pp. 72-74).

But the basic reason children give up autonomous action, Bettelheim thinks, is because they somehow get the feeling, correctly or incorrectly, that their existence is not desired. They feel that any action or willfulness—even any fussing—might be the last straw that brings about their destruction. In this respect, Bettelheim speculates, autistic children feel something like the prisoners in concentration camps for whom any action risked death. Thus, autistic children conclude that it is too dangerous to be independently assertive. Through a monumental act of will, they decide to do nothing and to be nothing, or to limit their actions to the small world that they can control (for example, endlessly spinning an ashtray, apparently oblivious to the happenings in the rest of the room). Moreover, since the sense of self depends on active interchange with the world, they forfeit the opportunity to build a self. Walling themselves off from the world, with no self inside, they live in an empty fortress (1967, pp. 73-83, 223).

It is sometimes thought that Bettelheim is blaming autism on the parents. Bettelheim says, however, that this is not his intention. Although he attributes the disorder to parent-child interactions, he thinks that inborn temperamental differences between parent and child may play a large role. For example, a fast, hyperactive baby boy may be out of tune with a slow-moving mother, and the boy will have difficulty finding appropriate feedback from his mother (1967, p. 29). In any case, Bettelheim thinks that the important thing is not what the parent does, but what the baby experiences—whether the baby can find the desired responses to his or her own actions (1967, p. 70).

THERAPY

Until fairly recently, autism was considered essentially untreatable. However, Bettelheim and a few others (especially Lovaas, 1973) have had some success. Bettelheim's treatment is on a residential basis; the children live in the Orthogenic School full time. The school, which houses 45 to 50 children with various disorders, has not attempted to treat more than six or eight autistic children at a time, but since the 1950s it has worked with 46 cases (Bettelheim, 1974, p. 50; 1967, p. 90). Treatment with the autistic children usually takes at least five years. Bettelheim's therapeutic principles emerge from his descriptions of three case studies (1967).

Love and Care

A crucial part of the school's environment is the provision of undivided love, care, and protection—a nurturance which probably counteracts any feelings by the children that others wish their destruction. The school assigns one and sometimes two counselors to each autistic child, and these counselors seem totally devoted to the care of their children. After a while, this loving care seems to register with the child. For example, Joey, a nine-year-old boy who had given up on people and had made himself into a machine, seemed to like being bathed,

> though for a long time this too was a mechanical procedure. In the tub he rocked hard, back and forth, with the regularity of an engine and without emotion, flooding the bathroom. If he stopped rocking, he did that too, like a machine.... Only once, after months of being carried to bed from his bath, did we catch a slight puzzled pleasure on his face and he said, in a very low voice, "They even carry you to your bed here." This was the first time we heard him use a personal pronoun (1967, p. 255).

Autonomy

Apart from this care, though, there is a sense in which Bettelheim believes that the staff cannot do things for the autistic children; for the most important thing that the children must develop is autonomy, and if autonomy is to be genuine, the children must gain it on their own. All the staff can do is create the right conditions of love and respect for the children and then hope that the children will begin to trust them sufficiently to take the first steps on their own.

The way in which love and care set the stage for autonomous action is illustrated by an incident with Laurie, a girl who came to the school at the age of seven years. Although Laurie was pretty and well dressed, she was completely inert and withdrawn. She had not uttered a word in more than four years. She also ate and drank very little and for months had kept her mough slightly open, which parched her lips. Laurie's counselor tried to

> wet her lips and also to oil them, to make her more comfortable. Her counselor rubbed her lips softly, and then gently put a finger in her mouth and on her tongue.... At first Laurie barely reacted, but later she seemed to like it, and for an instant she touched the finger with her tongue, may even have licked it for a moment (1967, p. 100).

Thus, the counselor's loving care seemed to inspire Laurie to take a small, but spontaneous initial action on her own.

Usually the children's first efforts at self-assertion occur around the issue of elimination, which is what happened, for example, in the case of Marcia. When Marcia came to the school at the age of 10½ years, she was completely unresponsive to people or objects, and she spoke only in single words that had some personal meaning to her alone. A central problem in her life was her constipation; she had stopped moving her bowels on her own after her mother had begun training her at two years. Since then, her parents had given her repeated enemas, an experience which, her doll-play later revealed, represented for her the feeling of being completely overpowered by huge adults. The staff's attitude toward her constipation illustrates the importance they put on the concept of autonomy. Bettelheim writes:

From the moment Marcia came we were convinced that if we were to force her to do anything, to give anything up, we could never help her out of her isolation. Nothing seemed more important than her acquiring the feeling that she was at least in charge of her own body. So from the beginning we assured her that we would not force her to move her bowels, that at the school she would never be given enemas, or any laxatives, and that in regard to elimination she could do as she wished (1967, p. 170).

She did not have to defecate in the toilet, "but could do it wherever and whenever it was easiest for her" (1967, p. 172). Soon Marcia began to soil, and the first place she defecated with any regularity was in the bathtub, the place where she seemed to feel the most relaxed and comfortable. After defecating in the tub, she frequently played with her stools. "After about a year and again in her own good time—though we occasionally made tentative suggestions—she began to eliminate in the toilet" (1967, p. 172).

The children's progress continues when they make initial attempts to relate to others. For example, after a year at the school, Marcia invited her counselor Inge to play a chasing game, exclaiming, "Chase!" However, Inge always had to maintain a certain distance, and Marcia never chased Inge in turn. Bettelheim speculates that through this game, in which Marcia was never caught, she was trying to master her feelings of being overpowered by adults. That is, she may have been trying to establish "through thousands of repetitions of the game that never again would anyone get hold of her and overpower the now barely emerging 'me' " (1967, p. 179). Thus, Bettelheim interprets Marcia's initial attempts to relate to others in terms of her need to establish autonomy. The counselors respected her wishes and always played the game on her own terms. Their attitude, in turn, seemed to win Marcia's trust to the point where she then tried new ways of relating to them.

The three cases suggest, finally, that progress in relating gives the children the courage to begin a new phase: Through symbolic play, they attempt to re-experience and master conflicts at the earliest developmental stage, the oral stage. For Marcia, it took some time before she could engage in purely oral play. For days, she repeatedly forced water in and out of a baby doll's mouth and rectum in exactly the same manner. Apparently, she first had to free herself of the death grip that enemas had on her total experience before she could work on feeding as a separate function (1967, p. 208). Finally, she separated the two activities by performing them in different rooms. When, however, her play did take on distinctively oral themes, she revealed that orality was fraught with its own grave dangers, as when, for example, she viciously beat a toy dog for daring to drink some milk (1967, p. 224). Marcia seemed to believe that oral intake was bad and could bring about the severest retaliation. Gradually, though, she was able to experiment with pleasurable ways of drinking in doll play and even to enjoy drinking itself.

The course of therapy, it should be noted, is not something that Bettelheim determines beforehand. The children take the lead in acting and exploring their problems, and the staff supports them the best they can. This is often difficult. Marcia's water play, for example, flooded the floors and required enormous work mopping up. But the staff members usually tolerate such behavior the best they can, for the children are trying to master their experience (1967, p. 204, 217).

Marcia eventually made a partial recovery. After five years in the school, she was talking to others and seemed capable of the full range of emotional expression. Her intellectual abilities, however, lagged behind; she was only reading at the fourth grade level. Perhaps she had entered the school at too late an age (10½ years) to permit a full recovery. Still, when she returned home, after seven years in the school, she was able to take care of herself and perform useful tasks. More important, she was no longer the frozen child she once had been.

Attitude toward Symptoms

One of the most radical aspects of Bettelheim's therapy is his attitude toward symptoms. For most mental health workers, symptoms (e.g., self-stimulation and peculiar gestures) are to be directly eliminated, or, at best, tolerated. Bettelheim, however, points out that the symptoms are what the children have spontaneously developed to gain some relief from, and even some mastery over, their tensions. They represent the child's greatest spontaneous achievement to date. Accordingly, they deserve our respect. If, instead, we disparage the symptoms—if, for example, we encourage the child to drop them—we cannot convey our respect for the child either (1967, p. 169).

The staff's attitude toward symptoms is illustrated by their approach to Marcia's behavior in the dining room. When Marcia first came to the school, she ate only candy, and in the dining room she plugged her ears with her forefingers and her nose with her little fingers, apparently to protect herself from something dangerous in the situation. This habit made it impossible for her to eat with her hands. The staff thought about telling her that it was O.K. to unplug her ears and nose, but they realized that this communication would fall woefully short. For if it were O.K. for her to unplug them she would do it herself. Similarly, they did not feel that an offer to feed her themselves would convince her that they understood her plight; if she could trust anyone to feed her, she would not need to plug herself up.

> Our solution was to offer to plug the ears for her; then she could have some fingers free to eat with. Hearing our offer, Marcia promptly plugged her nose with her forefingers and with her other fingers brought food to her mouth by bending as close to the plate as she could—a performance that astonished both the other children and all adults present (1967, pp. 169-70).

Many professionals would consider the staff's approach completely wrong; what they did was reinforce the psychotic behavior. However, they were trying to show a respect for the child's own devices for handling frightening feelings.

Sometimes the autistic symptoms include self-destructive behavior; the children try to hurt and damage themselves. In these cases, the staff does step in; they must protect the children (e.g., 1967, p. 268). However, other symptoms are respected as far as possible, for they are the children's autonomous constructions.

Phenomenology

Bettelheim's work has a phenomenological orientation. As a philosophy, phenomenology is exceedingly complex, but in psychology it generally means suspending our preconception that others think in some customary way and trying to enter into the other's unique world from the inside. It means putting oneself in the other's shoes (Ellenberger, 1958).

Bettelheim's phenomenological approach is illustrated by his attitude toward Marcia's plugging of her ears and nose. Although *he* knew it was O.K. for Marcia to unplug herself, he guessed that this was not *Marcia's* experience. Thus, he tried to see the world in terms of Marcia's unique inner experience and to act accordingly.

Bettelheim believes that autistic children will never leave their defensive positions as long as others are simply interested in getting them to see the world as they (adults) see it.

> This is exactly what the psychotic child cannot do. Instead, our task . . . is to create for him a world that is totally different from the one he abandoned in despair, and moreover a world he can enter right now, as he is. This means, above all, that he must feel we are with him in his private world and not that he is once more repeating the experience that "everyone wants me to come out of my world and enter his" (1967, p. 10).

The phenomenological task is especially difficult in the case of autistic children for two reasons. First, socialized adults who are fairly well adjusted to the external world are not readily in tune with the horror with which the autistic child regards it. The phenomenological task is difficult, secondly, because autistic children in many ways have never transcended the experiential modalities of the infant—modalities which are largely preverbal and foreign to us as adults

Summary

Bettelheim's therapy, then, includes (a) a great deal of love and care for the autistic children, which (b) enables them to trust others to the point where they will dare to take steps toward autonomous action. Bettelheim also believes (c) that the children will only make progress if they are given full respect as human beings, including respect for their symptoms as their greatest efforts to date to relieve their suffering. Finally (d), they will move out of the autistic position only if the staff somehow communicates to them that it does not simply want the children to enter its own world, but that it is trying to understand the children's own unique experience.

EVALUATION

Bettelheim has tried to evaluate the success of his treatment (1967, pp. 414-16). He believes that the Orthogenic School has had success with four-fifths of the autistic children. That is, about this number have been able to make at least a fairly good adjustment in society, including meaningful relationships with others. Of these, about half, for all intents and purposes, have been cured; although they may still show some personality quirks, they are doing well in their studies or are earning a living on their own. It is difficult to evaluate such statistics, however, because there are no reports on the reliability of these judgments. We do not know, that is, if neutral observers would agree with Bettelheim's assessments. Nevertheless, Bettelheim's case reports and films indicate that the children have made substantial progress, so we can have some confidence in his impressions.

Bettelheim's success is, for our purposes, very important because he takes a distinctly developmental approach and shows that it can work. Bettelheim, unlike many other workers in the mental health field, does not actively try to change the children's behavior. His staff is, in a sense, more passive. The staff assumes that if they set the right conditions, of love and care, the children will begin to take the steps toward health on their own. It is the children who take the lead. Bettelheim, as a theorist in the developmental tradition, believes that even in these very disturbed children there are inner forces toward growth and autonomy that will emerge in the right environment.

Rousseau, Gesell, and other developmentalists distinguish between autonomous growth and socialization. If we actively try to change children's behavior, we usually become socializers. We adopt socially appropriate behavior as our goal and try to teach children accordingly. The behaviorist Lovaas (1973), for example, tries to reinforce socially appropriate behavior, such as language, while eliminating socially inappropriate behavior, such as peculiar psychotic gestures. Bettelheim, in contrast, puts such a high premium on autonomous development that he frequently tolerates socially "deviant" acts. We have seen, for example, how he respects many psychotic symptoms, for they are the child's autonomous creations. We also have seen how the staff permitted Marcia to move her bowels in the bathtub and to flood the floors. They did so because she was taking steps on her own and exploring her problems. Bettelheim says that "too often children's progress is viewed not in terms of a move toward autonomy but of the convenience of a society that cares less about autonomy than conformity, and of parents who prefer not to clean their children's underclothes, no matter what" (1967, p. 294). The real question, he says, is "when, and for what gains, we ought to strip away social adjustment for the sake of personal development" (1967, p. 293).

Bettelheim, then, provides valuable support for the developmentalist position, which tries to let the children take the lead and do their own growing. His treatment is not the only one that is effective; Lovaas, in particular, has had fairly equal success with behavioral techniques. But Bettelheim gives us a reason to consider the developmentalist alternative.

I have implied that Bettelheim's approach differs not only from the behaviorists' but from that of most mental health workers. At the same time, there are a number of child psychoanalysts who would substantially agree with his approach. Most child analysts, of course, work with less disturbed children and therefore do not need to become active caretakers. Nevertheless, they often share Bettelheim's developmentalist orientation. That is, they do not try to get children to behave in the "correct" ways, but try to create a climate of acceptance and understanding that will enable children to take the initiative in exploring their problems.

The major weakness of Bettelheim's work is a lack of respect for standard scientific methodology. As mentioned, we would like more reliable outcome data. We also would like information on the relative effectiveness of Bettelheim's specific procedures, as indicated by objective measures. Bettelheim does not even attempt these. Nevertheless, Bettelheim's work and moving case reports are a significant contribution to the psychoanalytic and developmental literature.

10

Schachtel
on Childhood Experiences

BIOGRAPHICAL INTRODUCTION

Erikson and Bettelheim are considered Freudians, albeit quite innovative Freudians. In this and the next chapter, we will consider two theorists who stood farther apart from the Freudian mainstream—Schachtel and Jung. Schachtel, in fact, was more of a cognitive theorist in the line of Rousseau, Montessori, and Piaget. He thought that a cognitive approach could enlarge and amplify some basic Freudian observations.

Ernest Schachtel (1903-1975) was born and grew up in Berlin. His father wanted him to become a lawyer, which he did, even though he was more interested in philosophy, sociology, and literature. Schachtel practiced law for eight years, until 1933, when the Nazis had him jailed and then sent to a concentration camp. After his release, he worked on family research in England and Switzerland and then came to New York in 1935, where he received psychoanalytic training. Schachtel worked as a psychoanalyst the rest of his life, with a special interest in Rorschach (inkblot) testing, and, even more, in problems of child development (Wilner, 1975).

BASIC CONCEPTS

Schachtel was most specifically interested in the problem of *infantile amnesia,* our inability to remember most of the events of our first five or six years of life (Schachtel, 1959, p. 286). This curious gap in our memories was first noted by Freud, who pointed out that as infants we had many intense experiences—loves, fears, angers, jealousies—and yet our recall for them is very fragmentary. Freud's explanation was that this amnesia is the product of repression. Early sexual and aggressive feelings became linked to shame, disgust, and loathing, and therefore were repressed into the unconscious (Freud, 1905, pp. 581-83).

Schachtel believed that Freud was partly right, but Schachtel also pointed out two problems with the repression hypothesis. First, childhood amnesia is quite pervasive; we forget not only the sexual and hostile feelings that we might have had cause to repress, but we forget almost every other aspect of our early childhoods as well. Second, even psychoanalytic patients, who sometimes can get beneath the repression barrier, are still unable to remember much of their first few years. Thus, childhood amnesia must have an additional source (Schachtel, 1959, p. 285).

Schachtel suggested that childhood amnesia primarily has to do with *perceptual modes of experience.* Most adult experience and memory is based on verbal categories. For example, we look at a painting and say to ourselves, "That is a picture by Picasso," and this is how we remember what we have seen (see Slobin, 1971, pp. 102-4). Childhood experience, in contrast, is largely preverbal. It is, as Rousseau said, more directly based on the senses. As a result, it cannot be tagged, labeled—and later recalled—through verbal codes, and it therefore is lost to us as we grow up.

Schachtel divided childhood experience into two stages: infancy and early childhood. Let us look at the modes of perception during these two stages and compare them with the orientations of the adult.

Infancy (Birth to One Year)

In infancy, we rely on certain senses in particular. One vital sense is *taste.* Babies, who take many things into their mouths, have more taste buds than adults do and probably can make rather fine discriminations through this modality (1959, p. 138, 300). Babies also experience objects and people by their *smells.* Since babies are often held on the lap, and since young children, when standing, reach up only to the adult's waist, they probably are quite sensitive to the smell of legs, lap, and the sexual and excretory organs of the adult (1959, p. 138). Infants, Schachtel said, know what the mother tastes and smells like long before they know what she looks like. They probably can tell when she is tense or calm through these senses (1959, p. 299). Babies also are very sensitive to *touch* and react, for example, to the state of the mother as revealed by her relaxed or tense hold. Babies, finally, react sensitively to temperature, through the *thermal sense* (1959, p. 92).

Schachtel called these senses *autocentric.* By this he meant that the sensations are felt in the body. When, for example, we taste or smell food, our sensations are felt in the mouth or in the nostrils. Similarly, the experiences of hot and cold, and of being held and touched, are felt in or on the body. The autocentric senses are distinguished from the predominantly *allocentric* senses, *hearing,* and, especially, *sight.* When we use these sensory modalities, our attention is directed to the outside. When, for example, we look at a tree, we usually focus outward, on the object itself (1959, pp. 81-84, 96-115).

The autocentric senses are intimately bound up with feelings of pleasure and unpleasure, of comfort and discomfort. Good food, for example, produces a feeling of pleasure; rancid food produces disgust. The allocentric senses, in contrast, are usually more neutral. We experience no keen pleasure or disgust, for instance, in looking at a tree. The baby's predominantly autocentric experience, then, is tied to the pleasure principle, as Freud said.

Adult memory categories—which are predominantly verbal—are poorly suited for the recall of autocentric experience. We have a fair number of words for describing what we see but very few words for describing what we smell, taste, or feel. "A wine," for example, "is said to be dry, sweet, robust, fine, full, and so on, but none of these words enables one to imagine the flavor and bouquet of the wine" (1959, pp. 298-99). Poets can sometimes create vivid images of visual scenes, but they are unable to do so with respect to smells and tastes. The world of the infant, then, which is so much a world of smells, tastes, and bodily sensations, is not subject to verbal codes and recall.

Schachtel called special attention to the sense of smell. Western cultures practically outlaw discriminations based on this sense. If, upon being introduced to a man, I were to go up and sniff him, I would be considered extremely uncouth (although it is perfectly acceptable to inspect him visually at a distance). In English to say, "He smells," is synonymous with saying, "He smells badly." We do, of course, use perfumes and are aware of some fragrances, but on the whole our discriminations based on this sense are very limited.

The taboo on smell is probably related to the fact that odor is the primary quality of fecal matter. Babies, as Freud noted, seem to enjoy the smell of their feces, but socializing agents teach them otherwise. The result is that children repress specific anal experiences, as Freud said. But it is more than this. Children quit making fine discriminations based on this sense altogether. Consequently, their early experiences are lost, for they do not fit into the accepted categories of experience.

Early Childhood (One to Five Years)

During infancy, we do not generally welcome changes in internal or external stimulation. Sudden changes—such as sharp hunger pains, shivering cold, or the loss of bodily support—can be quite threatening. The infant would, as Freud said, like to remain embedded in a warm, peaceful, protective environment, much like the womb (Schachtel, 1959, pp. 26, 44-68).

At about one year of age, however, the child's basic orientation undergoes a change. Children become relatively less concerned about their security; under maturational urging, they take a much more active and persistent interest in new things. To some extent, they still rely on the autocentric senses, as when, for example, they put objects into their mouths. But they now increasingly utilize the pure allocentric senses—hearing, and especially, sight. They examine and explore new objects by looking at them.

The young child's attitude is one of openness to the world. The child has the capacity to take everything in, no matter how small or insignificant to us, in a fresh, naive, and spellbound manner. A little girl, coming across an insect, will stop and examine it intently. To her, the insect is full of new and fascinating possibilities. She perceives each new object with a sense of wonder and awe.

This openness contrasts markedly with the predominant attitude of adults. Most adults simply label objects—for example, "That is an ant"—and then go on to something else. Adults use the same allocentric senses—sight and hearing—but not in a fully allocentric way, not with an openness to things themselves. As adults, our greatest need is not to explore the world in all its richness but to reassure ourselves that everything is familiar, as accustomed and expected.

It is not easy to understand why adults are in such a hurry to name, label, and classify things, but the answer probably has to do with the process of socialization. As children grow up, they find that grown-ups and peers have standard, conventional ways of describing the world, and the pressure is great to adopt them. The older child and the adult become afraid of looking at things in any way that might be different from others. There is always the threat of feeling odd, ignorant, or alone. Just as infants need the security of caretakers, adults need the security of belonging and conforming to their cultural group. Consequently, they come to see what others see, to feel what everyone feels, and to refer to all experiences with the same clichés (1959, pp. 204-6, 176-77). They then think that they know all the answers, but they really only know their way around the conventional pattern, in which everything is familiar and nothing a cause for wonder (1959, p. 292).

What the adult becomes capable of remembering, then, is that which fits into conventional categories. For example, when we take a trip, we see all the sights we are supposed to see, so we can be sure to remember exactly what everybody else remembers too. We can tell our friends that we saw the Grand

Canyon, that we stopped at six Howard Johnson restaurants, that the desert looked beautiful at sunset (just like in the postcards), and the routes we took. We cannot, however, give any real idea of what the country was like. The trip has become a mere collection of clichés (1959, p. 288).

Similarly, our journey through life is remembered in terms of conventional signposts. A man can tell us about his birthdays, his wedding day, his jobs, the number of children he had and their positions in life, and the recognitions he received. But he will be unable to tell us about the truly special qualities of his wife, of his job, or of his children, for to do so would mean opening himself to experiences that transcend the conventional categories of perception (1959, p. 299).

Among adults, Schachtel said, it is primarily the sensitive artist who has retained the young child's capacity to view the world freshly, vividly, and openly. Only the exceptional artist can experience things with the young child's sense of wonder at watching an insect walk, at the way a ball squeezes, bounces, and responds to the hand, or at the way water feels and looks as it is poured. For most of us, unfortunately, "the age of discovery, early childhood, is buried deep under the age of routine familiarity, adulthood" (1959, p. 294).

In summary, then, neither the autocentric experiences of the infant nor the allocentric experiences of the young child fit into the adult's way of categorizing and remembering events. The infant's world of tastes, smells, and touch, as well as the young child's fresh and open experience of things in all their fulness, are foreign to the adult and are not subject to recall.

IMPLICATIONS FOR EDUCATION

Most of Schachtel's thoughts on child-rearing and education concerned the child as he or she begins actively to explore the world. Schachtel wanted us to preserve and encourage the young child's bold curiosity. Unfortunately, we usually stifle it.

For example, when babies begin to handle and examine everything they see, parents often become overly anxious. Parents, as Montessori observed, are afraid that their children are acting too rough, that they might break things, or that they might hurt themselves. Actually, it is usually simple enough to "child proof" a house—to remove all breakable and dangerous objects—and then permit the child to explore. Nevertheless, parents often become anxious at this point, and the result is that children learn that it is dangerous to become too curious about the world (Schachtel, 1959, p. 154).

Adults also may discourage children's curiosity by the way they name, label, and explain things to children. For example, when a child becomes curious about something, the adult often simply tells the child the object's name, implying that there is nothing more to know about it (1959, p. 187). If a little girl shouts, "Da!" and points excitedly at a dog, the father says, "Yes, that's a dog," and then urges her to continue on with their walk. He teaches her that the conventional category, the word, "explains" the object. Instead, he might say, "Yes, that's a dog!" and stop and observe it with her. In this way, he respects and encourages her active interest in the world.

Schachtel, on the whole, told us more about how parents, teachers, and peers stifle the child's curiosity than how we might protect and encourage it. Like Rousseau, he implied that the most important thing is to avoid negative influences. If we can lessen our tendency to close off their world, children themselves will take an open, active interest in it, from their own spontaneous tendencies.

EVALUATION

Freud thought that the great tragedy of life was that, in order for us to live in society, we repress so much of ourselves. Freud had in mind the repression of instinctual drives. Erikson elaborates on this theme, suggesting that positive potentials for autonomy, initiative, and other strengths usually become somewhat curtailed in the course of socialization. Schachtel's contribution is an enlarged conception of just how much we do lose. It is not just that we repress our drives, or even that ego strengths such as autonomy or initiative are restricted, but that we sacrifice entire modes of experience. The baby who is in direct contact with objects through the senses of smell, taste, and touch and the child who is openly curious about the world grow into well-socialized adults who view the world through very narrow, verbal, conventional schemes. Perhaps Schachtel, by helping us appreciate the extent of the problem, also puts us in a position to begin to correct it.

Schachtel also demonstrated, in the course of his writings, the value of a phenomenological approach to childhood. He provided glimpses of how the infant's world might appear from the inside—which senses are most central and how experience is distinctive when these senses are dominant. The phenomenological approach deserves wider application in developmental psychology.

At the same time, Schachtel's work suffered from certain oversimplifications. For one thing, he probably underestimated the importance of vision in infancy. For example, we saw in our chapter on ethology how the baby's attention to visual patterns, such as the face, is very important. Schachtel also gave a one-sided picture of language. He felt that the acquisition of language stifles creativity, for the child learns to substitute the conventional label or word for the experience itself. While this may be true, the child's mastery of language also is a creative process, as we shall see in our chapter on Chomsky.

Still, Schachtel did much to keep psychologists and psychoanalysts mindful of the value of the radical Rousseau-like way of thinking. He pointed out how very different the child's world is from ours and how much human potential for fresh creative experience may be lost in the process of becoming a well-adjusted member of the conventional social order.

11

Jung's
Theory of Adulthood

BIOGRAPHICAL INTRODUCTION

We have noted that few theorists have concerned themselves with development during the adult years. Erikson is one notable exception. Another was C. G. Jung (1875-1961), whose psychoanalytic theory dealt primarily with the issues of adulthood and aging.

Jung was born in Kesswil, a village in northeastern Switzerland. His childhood was mostly unhappy. He experienced the tensions of his parents' marital difficulties and was usually quite lonely. School bored him and even precipitated fainting spells when he was 12 years old (Jung, 1961, p. 30). Because his father was a pastor, Jung went to church, but he disliked it and got into bitter religious arguments with his father. Jung's primary enjoyments during childhood and adolescence were exploring nature and reading books of his own choosing— drama, poetry, history, and philosophy.

Despite his problems, Jung did well in high school and went on to earn a medical degree. He then began practicing psychiatry in Zurich, where he quickly developed a lasting interest in psychotic disorders. Jung's work—including his invention of the word-association test—suggested the importance of Freud's ideas. His colleagues warned him that any alignment with Freud would jeopardize his career, but he went ahead and indicated the importance of Freud anyway (Jung, 1961, p. 148). Freud, of course, appreciated Jung's support, and when the two men met they found they had much in common. For some time, Freud treated Jung like a son and chosen disciple. Jung, however, disagreed with aspects of Freud's theory, particularly with the attempt to reduce all unconscious events to sexual drives. Jung believed the unconscious contains many kinds of strivings, including religious and spiritual ones. In 1912 Jung decided to develop his own ideas, and in 1913 the two men severed ties.

After parting with Freud, Jung lost his footing. He began having uncanny, deeply symbolic dreams and experienced frightening visions during his waking hours. For example, in one vision he saw the Alps grow and then saw mighty yellow waves drown thousands of people and turn into a sea of blood. Since World War I broke out the next year, he believed that his vision carried a message which related to events far beyond himself (Jung, 1961, pp. 175-76).

Jung realized that he was on the brink of psychosis, but he nevertheless decided to submit to the unconscious—to whatever was welling up and calling him from within. It was his only chance of understanding what was happening to him. Thus he embarked on a terrifying inner journey in which he frequently felt himself descending into lower and lower depths. At each region, he saw archaic symbols and images, and he communicated with demons, ghosts, and strange figures from the distant historical past. During this period, his family and professional practice served as bases of support in the outer world. Otherwise, he was certain, the images welling up from within would have driven him completely out of his mind (1961, p. 189).

Gradually, after about four years, he began to find the goal of his inner quest. This happened when he increasingly found himself drawing geometrical figures, symbols composed of circles and squares which he would later call mandalas (see Figure 11.1). The drawings represented some basic unity or wholeness, a path to the center of his being. Jung said that during his psychotic state

> I had to let myself be carried along by the current, without knowing where it would lead me. When I began drawing mandalas, however, I saw that everything, all the paths I had been following, all the steps I had taken, were leading back to a single point—namely to the midpoint. It became increasingly clear to me that the mandala is the center (1961, p. 196).

Jung thus began to understand his break as a necessary inner journey which led to a new personal integration. Support for this, however, came some eight years later, when he dreamed of a curiously Chinese-looking mandala, and the next year he received a book in Chinese philosophy which discussed the mandala as the expression of the unity of life. Jung then believed that his own experience partook of an unconscious universal quest for psychic wholeness.

FIGURE 11.1   A Tibetan mandala.

Reprinted from *The Collected Works of C. G. Jung,* trans.
R. F. C. Hull, Bollingen Series XX. Vol. 9, I, *The Archetypes
and the Collective Unconscious,* copyright © 1959, 1969 by
Princeton University Press. Figure 1, "Concerning Mandala
Symbolism" reproduced by permission of the Jung Estate
and Princeton University Press.

Jung made the exploration of the unconscious and its symbols the center of his research for the rest of his life. He continually explored his own dreams and fantasies and those of his patients. He also extensively studied the myths and art of numerous cultures, finding in these productions the expression of universal, unconscious yearnings and tensions.

PERSONALITY STRUCTURE

Although Jung was most concerned with the nature of the unconscious, he developed a theory of personality which encompasses various systems of personality functioning. We will first describe Jung's ideas on personality structure, and then discuss his views on how the personality develops.

The ego. The ego is roughly equivalent to consciousness. It includes our awareness of the external world, as well as our consciousness of ourselves (Jung, 1933, p. 98; Whitmont and Kaufmann, 1973, p. 93).

The persona. The persona is the ego's mask, the image one presents to the outer world. Our personas vary with our roles. For example, a man presents one image to his business associates, another to his children. Some people develop their personas to the exclusion of deeper parts of the personality. At some point they, or others, sense that there is little of substance beneath the superficial front (Jung, 1961, p. 385). It also is true, however, that we need this part of the personality to deal effectively with others. It is often necessary, for example, to convey an image of confidence and decisiveness if we want others to listen to us (Jacobi, 1965, p. 37). To the extent that the personality is balanced, the persona will be developed, but not to the exclusion of other parts.

The shadow. The shadow consists of those traits and feelings that we cannot admit to ourselves. It is the opposite of our ego or self-image; it is the Mr. Hyde of Dr. Jekyll. In dreams, the shadow is projected onto people of the same sex, as when a man dreams about an evil killer, or when a woman dreams about an uncouth harlot. In our daily lives, our shadow often shows when we are in awkward situations and, for example, suddenly blurt out a hostile remark that "doesn't sound like me at all." We also see the projections of our shadows when we complain about "the one thing I cannot stand in people"; for such vehemence suggests that we are really defending against an awareness of this quality in ourselves (von Franz, 1964, p. 174).

In most cases, the shadow is largely negative, for it is the opposite of our positive self-image. However, to the extent that our conscious self-image contains negative elements, the unconscious shadow will be positive (Jung, 1961, p. 387). For example, a young woman who considers herself unattractive may dream about a beautiful lady. She considers this lady somebody else, but it may really represent her own beauty wishing to emerge. Whether the shadow is positive or negative, it is important to get in touch with it. Insight into the nature of one's shadow is the first step toward self-awareness and the integration of the personality (Jung, 1933, p. 33).

The anima and animus. Chinese Taoists speak of the Yin and the Yang, the feminine and the masculine sides of our personalities. According to Jung, the feminine principle includes capacities for nurturance, feeling, and art, and a oneness with nature. The masculine principle includes logical thinking, heroic assertion, and the conquest of nature (Jung, 1961, pp. 379-80; Whitmont and Kaufmann, 1973, p. 94). We are all biologically bisexual, and we all identify with people of both sexes, so we all possess both masculine and feminine traits. However, there are also genetic sex differences, which socialization pressures exaggerate, forcing women to overdevelop their feminine side and men to over-emphasize their masculine nature. The result is that the "other side" is repressed and weak. Men tend to become one-sidedly independent, aggressive, and intellectual; they neglect capacities for nurturance and relatedness to others. Women develop their nurturant and feeling sides but neglect their capacities for self-assertion and logical thought. Nevertheless, the neglected aspects do not disappear but remain active and call out to us from the unconscious. In men, the feminine side emerges in dreams and fantasies as "the woman within," the anima. In women, the "man within" is called the animus (Jung, 1961, p. 380).

The personal unconscious. Jung thought that the unconscious consists of two layers. The first is the personal unconscious, which contains all the tendencies and feelings that we have repressed during our lifetime (Jung, 1961, p. 389). Much of the shadow is in the personal unconscious. It might include, for example, a man's hostile feelings toward his father which, as a child, he needed to repress. The anima and animus also are partly, but not completely, found in this unconscious region. For example, a woman may have repressed her experiences of her father as seductive, experiences which then contribute to her animus and reside in her personal unconscious.

The collective unconscious. Each individual's personal unconscious is unique, for each person has repressed different thoughts and feelings during his or her lifetime. However, Jung also believed that there exists, at the deepest layer of the psyche, a collective unconscious which is inherited and shared by all mankind. The collective unconscious is made up of innate energy forces and organizing tendencies called *archetypes.* We can never know the archetypes directly, but we can learn about them through archetypal images found in the myths, art, dreams, and fantasies of peoples throughout the world. Through these images, people try to express their deepest inner yearnings and unconscious tendencies. They include the image of the Earth Mother, the Wise Old Man, rebirth, death, the trickster, the witch, and God. The archetypes also influence the nature and growth of the other parts of the personality. For example, a woman's animus results not only from her experiences with her father and other men but from an unconscious, universal tendency to find some Wise Man or Good Father.

Different cultures express archetypal themes in somewhat different ways, but humans everywhere have always been fascinated and impressed by them. Jung said,

> The concept of the archetype . . . is derived from the repeated observation that, for instance, the myths and fairytales of world literature contain definite motifs which crop up everywhere. We meet these same motifs in the fantasies, dreams, deliria, and delusions of individuals living today . . . They have their origin in the archetype, which in itself is an irrepresentable unconscious, pre-existent form that seems to be part of the inherited structure of the psyche and can therefore manifest itself spontaneously anywhere, at any time (1961, p. 380).

Although Jung said archetypes are essentially unknowable, he also likened them to instincts in animals—for example, a bird's innate schema of the parent (1964, p. 58). Perhaps archetypes also can be likened to the innate perceptual tendencies Gestalt psychologists talk about (R. Watson, 1968, Ch. 19; Arnheim, 1954). For example, we may have an inner sense of what constitutes a harmonious form. Mandalas probably strike us as beautiful because they correspond to our senses of proportion, balance, and good organization.

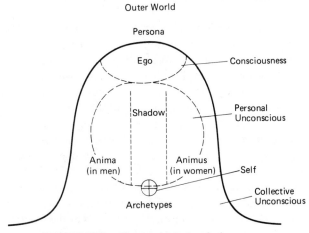

FIGURE 11.2   The psyche in Jung's theory.

The Self. The most important archetype is that of the Self, our unconscious striving for centeredness, wholeness, and meaning (Jung, 1961, p. 386). The Self is an inner urge to balance and reconcile the opposing aspects of our personalities. It is represented throughout the world in drawings of mandalas, figures in which all sides are perfectly balanced around a center point. The Self is also expressed by our search for God, the symbol of wholeness and ultimate meaning (Jung, 1961, p. 382).

Introversion and extraversion. Although the Self is the ultimate goal in life, no one ever fully attains it. We all develop in one-sided ways. Most of us, for example, develop our consciousness and neglect our unconscious lives. Women neglect their masculine side, men their feminine side. Jung developed other concepts to describe opposite tendencies, of which we develop one to the exclusion of the other. One such polarity is introversion-extraversion. The extravert confidently engages in direct action; the introvert hesitates and reflects upon what may happen. For example, at a party an extraverted young woman immediately walks over to others and strikes up a conversation with them. The introverted woman hesitates; she is caught up in her inner state, her fears, hopes, and feelings. The extravert moves outward, toward the world; the introvert is more secure in his or her inner world, and takes more pleasure in activities such as reading and art. We all have both tendencies, but are predisposed toward one, leaving the other underdeveloped and unconscious (Jung, 1945).

THEORY OF DEVELOPMENT

The First Half of Life

The personality develops along different lines during the first and second halves of the life cycle. The first period, until the age of 35 or 40, is a time of outward expansion. Maturational forces direct the growth of the ego and the unfolding of capacities for dealing with the external world. Young people learn to get along with others and try to win as many of society's rewards as possible. They establish careers and families, and they do what they can to advance up the social ladder of success. To do so, it is usually necessary for women to develop their feminine traits and skills, and men their masculine ones.

During this phase, a certain degree of one-sidedness is necessary and even valuable; for the young person needs to dedicate himself or herself to the task of mastering the outer world. It is not especially advantageous for young people to be too preoccupied with their self-doubts, fantasies, and inner natures (Jung, 1933, p. 109). For the task is to meet the demands of the external environment confidently and assertively. As one can imagine, extraverts, rather than introverts, have an easier time of it during this period (Jacobi, 1965, p. 42).

The Mid-Life Crisis

About age 40, the psyche begins to undergo a transformation. The individual feels that the goals and ambitions which once seemed so eternal have lost their meaning. Quite often, the person feels depressed, stagnant, and incomplete, as if something crucial is missing. Jung observed that this happens even among people who have achieved a good measure of social success. For, ". . . the achievements which society rewards are won at the cost of a diminution of personality. Many—far too many—aspects of life which should have been experienced lie in the lumber-room among dusty memories" (1933, p. 104).

The psyche itself provides the way out of this crisis. It urges the person to turn inward and examine the meaning of his or her life. This turning inward is prompted by the unconscious, the region in which all the repressed and unlived aspects of the self have grown and now clamor to be heard. The unconscious calls out for recognition in order to bring about psychic balance and harmony (Jung, 1933, pp. 17-18, 62-63).

The unconscious speaks to us primarily through dreams. For example, early in his analysis, a man who was depressed and felt that his life was meaningless had the following dream:

> I am standing totally perplexed, in the midst of a Casbah-like city with serpentine and winding small streets, not knowing where to turn. Suddenly I see a young, mysterious woman whom I had never seen before, pointing with her hand the direction out. It had a very awesome quality to it (Whitmont and Kaufmann, 1973, p. 95).

The dream's message is that the man must pay attention to his anima (the mysterious young woman) for the direction out of his impasse. For it is this side of himself that he has so far failed to appreciate and develop.

As adults examine their lives and listen to unconscious messages, they sooner or later encounter images of the Self, the symbols of wholeness and centeredness. For example, a middle-aged man, a highly successful executive who had been increasingly experiencing tension and a suicidal depression, had a dream in which a water devil attacked him and maneuvered him to the edge of an abyss. The creature then rescued him and gave him a drill, which the dreamer was supposed to use to dig to the center of the earth. In Jungian terms, the dream's key figure, the water devil, is initially evil, for it represents elements in the unconscious which the conscious ego has considered inadmissable—destructive urges. Yet the devil turns into a helper, indicating that the man must confront the negative aspects of himself if he is ever to become whole and find his true center (Whitmont, 1969, pp. 224-25).

As the dream continued, the man found himself in a secret chamber where a meeting was in progress around a square table. Presiding were a splendid knight and a Lord Mayor. However, the dreamer was led to the table by a delinquent boy and a dirty, tramp-like man who looked like a friar. They sat at the table and gave him some repulsive food which he ate anyway. He then pushed his drill into the ground, and flowers grew around the drill, and the drill was transformed into a blossoming tree.

According to a Jungian analysis, the splendid knight and the Mayor represent the heroic and authoritative aspects of the personality which the man has consciously realized as a business administrator. The delinquent and the ragged friar stand for the neglected, devalued aspects of the self, elements he has so far repressed. In particular, this man had regarded his religious inclinations as soft and escapist, as tendencies which would undermine his manhood. But it is the monk and the delinquent—repressed shadow figures—who lead him to the square table, the symbol of psychic unity. There he eats repulsive food—for no one likes looking into the repressed parts of the personality—but in so doing he sees that new growth (the blossoming tree) will emerge (Whitmont, 1969, pp. 224-27).

You will notice that the Jungian approach to dreams differs from that of the Freudians. Freud considered dreams to be the end-products of distortion and disguise. Jung, in contrast, believed that dreams express unconscious meanings fairly directly (1933, p. 13). In the dream cited above, the delinquent and the monk are interpreted quite literally, as symbols for the delinquent and religious aspects of the personality. Further, the unconscious, the source of dreams, is not just a seething pit of base impulses and desires, as it was for Freud. Rather, the unconscious, as a part of nature, can be a creative force, directing us out of our current stalemates, as the above dreams illustrate. Dreams tell us which aspects of the self we have neglected and must get in touch with (Jung, 1933, pp. 62-63). This does not mean that we should actually live out our evil tendencies. However, it does mean that we should learn about them, so we can control them, rather than the other way around (Whitmont, 1969, pp. 227-30).

The road toward health and growth—toward the unattainable goal of the Self—is called *individuation* (Jung, 1933, p. 26). Individuation involves not only achieving a measure of psychic balance but separating ourselves from our ordinary conformity to the goals and values of the mass culture. It means finding one's individual way. Each person's true nature partakes of universal archetypes, but it is also based on unique experiences and potentials which must be discovered (Jung, 1961, p. 383; Jacobi, 1965, pp. 83-87).

The middle of life, then, is marked by a transformation in the psyche. We are prompted to begin turning our energy away from the mastery of the external world and to begin focusing on our inner selves. We feel inner urgings to listen to the unconscious to learn about the potentials we have so far left unrealized. We begin to raise questions about the meaning of our lives, which, after all, are now half over.

Although the focus from middle age onward becomes increasingly inward, the middle-aged adult still has the energy and resources for making changes in his or her external situation. In middle age, adults quite often take up long-neglected projects and interests and even make seemingly incomprehensible career changes. Men and women, Jung observed (1933, p. 108), begin giving expression to their opposite sexual sides. Men become less aggressively ambitious and become more concerned with interpersonal relationships. As an older colleague once told me, "As you get older you find that achievement counts for less and friendship counts for far more." Women, on the other hand, become more aggressive and independent. If, for example, a man loses interest in the family business, the wife may willingly take it over. The woman, like the man, is becoming increasingly inner-directed as she ages, but her enthusiasm for aggressive pursuits may temporarily counterbalance or stall this general inner orientation.

The changes of life in middle age can create marital problems. The wife may tire of her husband's intellectual condescension, for she is now trying to develop her own thinking side. The husband, on the other hand, may feel oppressed by his wife's tendency to treat him emotionally like a child. He no longer simply wants to be calmed and pampered for his moodiness, but he wants to explore the realm of feelings and relationships in a more mature manner. Changes such as these can disturb the marital equilibrium (Jung, 1931).

Although growth during the second half of life creates tensions and difficulties, the greatest failures come when adults cling to the goals and values of the first half of life (Jung, 1933, p. 109). For example, a middle-aged woman may desperately try to maintain the physical attractiveness of her youth. Or the middle-aged man may talk incessantly about his past athletic glories. In such cases, adults miss out on further development, which can emerge only when they risk a confrontation with the neglected parts of themselves.

Old Age

Jung said that "with increasing age, contemplation, and reflection, the inner images naturally play an ever greater part in man's life" (1961, p. 320). "In old age one begins to let memories unroll before the mind's eye . . ." (1961, p. 320). The old person tries to understand the nature of life in the face of death (1961, p. 309).

Jung believed that we cannot face death in a healthy way unless we have some image of the hereafter. If "I live in a house which I know will fall about my head within the next two weeks, all my vital functions will be impaired by this thought; but if on the contrary I feel myself to be safe, I can dwell there in a normal and comfortable way" (1933, p. 112). When Jung recommended that the aged entertain thoughts of an afterlife, he did not feel that he was simply prescribing some artificial tranquilizer. For the unconscious itself has an archetype of eternity which wells up from within as death nears.

Jung could not say, of course, whether our archetypal image of the hereafter is valid, but he believed that it is a vital part of psychic functioning and therefore tried to get some picture of it. He based his picture on his own last dreams and those of others near death. The archetypal image of eternity, in Jung's view, is not of some hedonistic paradise. Instead, he pictured the souls of the dead to be like a spellbound audience listening to a lecture, seeking information from the newly deceased on the meaning of life. Apparently they know only what they knew at the moment of death. "Hence their endeavor to penetrate into life in order to share in the knowledge of men. I frequently have the feeling that they are standing directly behind us, waiting to hear what answer we will give them, and what answer to destiny" (1961, p. 308). They continue to strive to attain in death that share of awareness which they failed to win in life.

In Jung's view, then, life after death is a continuation of life itself. The dead, like the aged, continue to struggle with the questions of existence; they wonder what it is that makes a life whole and gives it meaning—they search, in Erikson's term, for integrity. Jung's view of the continuation of this questioning fits well with his theory of the unconscious as something which extends beyond any finite life and participates in the tensions of the universe.

PRACTICAL IMPLICATIONS

As with the Freudians, Jung's theory is inseparable from its practical applications. Much of Jung's theory came out of his work with patients, who helped him understand the nature of the unconscious and the kinds of experiences that were necessary to integrate their lives. Jungian psychoanalysis is most applicable to adults and older people. In fact, over two-thirds of Jung's own clients were in the second half of their lives (Whitmont and Kaufmann, 1973, p. 110).

Jung's ideas also would seem of value even to those who never find their way to the analyst's office. Most adults, at one time or another, probably experience the special problems that come with growing older. Some knowledge of the kinds of changes that typically occur may help them get their bearings. For this reason, adults seem to derive benefits from a book such as *Passages* (Sheehy, 1976), which discusses the expectable crises of adulthood. Since this book popularizes ideas of Levinson (1977, 1978), and Levinson, in turn, owes much to Jung, one ultimately may find the deepest rewards from Jung himself.

Jung's relevance for psychology and psychiatry extends beyond his special insights into adult development. Because he considered religious questions vital and meaningful in their own right, ministers, priests, and others, who themselves must often work with emotionally distressed people, have found that Jung provides a valuable bridge between the religious and psychiatric professions.

Jung's writings, furthermore, anticipated some new thinking on the nature of psychotic disorders, especially the thinking of Laing (1967). Laing argues that it is wrong to view psychotic experiences as simply abnormal and bizarre. This attitude characterizes technological cultures which refuse to admit the validity of the inner world and, instead, make outer adjustment the sole objective. Laing contends that the psychotic experience, for all its pain, can be a meaningful inner voyage and a healing process. In this voyage, the therapist can serve as a guide, helping the patient understand his or her inner symbols. Jung's view was somewhat similar, and a knowledge of Jung would seem essential for anyone who wishes to understand psychosis.

EVALUATION

Jung occupies an unusual place in contemporary psychology. On the one hand, many of his ideas are too mystical for most psychologists. He not only posited a collective unconscious, but he believed in ESP and related phenomena (e.g., Jung, 1961, p. 190). Sometimes, in addition, he seemed unnecessarily determined to keep his concepts shrouded in mystery. For example, he said that the archetypes are unknowable, yet he compared them to instincts in animals —a topic certainly open to scientific investigation.

On the other hand, psychologists in general, and developmentalists in particular, are increasingly recognizing the importance of Jung's ideas. Developmentalists, as we have seen, have long been concerned with how we seem to lose so much of ourselves and our potentials as we become socialized, as we become adjusted to the external world. Jung agreed that this happens, but he saw new opportunities for individual growth in the adult years.

Moreover, a growing body of empirical research is lending support to Jung's ideas. For example, in a major study of adult men, Levinson (1977, 1978) interprets his findings in a Jungian context. Levinson found that the vast majority of his subjects underwent a crisis at about age 40 or 45, during which they began to experience "internal voices that have been silent or muted for years and now clamor to be heard" (1977, p. 108). Levinson concludes that the life structure of the 30s necessarily gives high priority to certain aspects of the self, those oriented toward social adjustment and achievement. But in the 40s, "the neglected parts of the self urgently seek expression and stimulate a man to reappraise his life" (1977, p. 108).

The impressive studies of Neugarten and her colleagues at the University of Chicago have also lent support to Jung's insights. Neugarten reports that for both sexes the 40s and 50s mark a "movement of energy away from an outer-world to an inner-world orientation" (1964, p. 201). Introspection, contemplation, and self-evaluation increasingly become characteristic forms of mental life (Neugarten, 1968, p. 140). Furthermore, men "become more receptive to their own affiliative, nurturant, and sensual promptings; while women seem to become more responsive toward, and less guilty about, their own aggressive, egocentric impulses" (Neugarten, 1964, p. 199). Thus, Neugarten's research, like Levinson's, documents some of the personality shifts Jung outlined.

These shifts, it is important to note, seem to occur consistently before external situations demand them (Havighurst *et al.,* 1968, p. 167). Adults seem to turn inward before they might be forced to do so by external losses, such as retirement, reduced income, or the loss of a spouse. Thus, there seems to be an intrinsic developmental process at work. Adults may have an inherent need to take stock, to resist the pressures of conventional roles, and to concern themselves with the neglected and unrealized aspects of the personality.

12

Learning Theory: Pavlov, Watson, and Skinner

In the preceding chapters, we have discussed theorists in the developmental tradition. These theorists believe that key developments are governed by internal forces—by biological maturation or by the individual's own structuring of experience. In this and the following chapter, we will describe the work of some of the theorists in the opposing, Lockean tradition—learning theorists who emphasize the processes by which behavior is formed from the outside, by the external environment.

PAVLOV AND CLASSICAL CONDITIONING

Biographical Introduction

The father of modern learning theory is Ivan Petrovich Pavlov (1849-1936). Pavlov was born in Ryazan, Russia, the son of a poor village priest. Pavlov himself planned to become a priest until the age of 21, when he decided he was more interested in a scientific career. For many years he devoted his attention to physiological investigations, and in 1904 he won the Nobel Prize for his work on the digestive system. It was just a little before this time, when Pavlov was 50 years old, that he began his famous work on conditioned reflexes. This new interest came about through an accidental discovery about the nature of salivation in dogs. Ordinarily dogs salivate when food touches their tongues; this is an innate reflex. But Pavlov noticed that his dogs also salivated *before* the food was in their mouths; they salivated when they saw the food coming, or even when they heard approaching footsteps. What had happened was that the reflex had become conditioned to new, formerly neutral stimuli.

For a while, Pavlov could not decide whether to pursue the implications of his new discovery, or to continue with his earlier research. Finally, after a long struggle with himself, he began studying the conditioning process. Still, Pavlov believed that he was working as a physiologist, not a psychologist. In fact, Pavlov required that everyone in his laboratory use only physiological terms. If his assistants were caught using psychological language—referring, for example, to a dog's feelings or knowledge—they were fined (R. Watson, 1968, pp. 408-12).

Basic Concepts

The classical conditioning paradigm. In a typical experiment (Pavlov, 1928, p. 104), a dog was placed in a restraining harness in a dark room and a light was turned on. After 30 seconds, some food was placed in the dog's mouth, eliciting the salivation reflex. This procedure was repeated several times—each time the presentation of food was paired with the light. After a while, the light, which initially had no relationship to salivation, elicited the response by itself. The dog had been conditioned to respond to the light.

In Pavlov's terms (1927, Lectures 2 and 3), the presentation of food was an *unconditioned stimulus* (US); Pavlov did not need to condition the animal to salivate to the food. The light, in contrast, was a *conditioned stimulus* (CS); its effect required conditioning.[1] Salivation to the food was called an *unconditioned reflex* (UR) whereas salivation to the light was called a *conditioned reflex* (CR). The process is called classical conditioning.

You might have noticed in this experiment that the CS appeared *before* the US; Pavlov turned on the light before he presented the food. One of the questions he asked was whether this is the best order for establishing conditioning. He and his students discovered that it is. It is very difficult to obtain conditioning when the CS follows the US (Pavlov, 1927, pp. 27-28). More recent studies have suggested that conditioning takes place most rapidly when the CS is presented about one-half second prior to the US (see Kimble, 1961).

Pavlov discovered several other principles of conditioning, some of which we will briefly describe.

[1] Pavlov actually used the terms "conditional" and "unconditional"; they were translated "conditioned" and "unconditioned," the terms psychologists now generally use.

Extinction. A conditioned stimulus, once established, does not continue to work forever. For example, Pavlov found that even though he could make a light a CS for salivation, if he flashed the light alone over several trials, the light began to lose its effect. Drops of saliva became fewer and fewer until there were none at all. At this point, extinction had occurred (Pavlov, 1928, p. 297).

Pavlov also discovered that although a conditioned reflex appears to be extinguished, it usually shows some *spontaneous recovery.* For example, in one experiment (1927, p. 58), a dog was trained to salivate to the mere sight of food—the CS. (Previously, the dog would salivate only when food was in its mouth.) Next, the CS alone was presented at three-minute intervals for six trials, and by the sixth trial, the dog no longer salivated. Thus, the response appeared to have been extinguished. However, after a two-hour break in the experiment, the presentation of the CS alone once again produced a moderate amount of salivation. Thus, the response showed some spontaneous recovery. If one were to continue to extinguish the response, without periodically re-pairing the CS to the US, the spontaneous recovery effect also would disappear.

Stimulus generalization. Although a reflex has been conditioned to only one stimulus, it is not just that particular stimulus that elicits it. The response seems to generalize over a range of similar stimuli without any further conditioning (Pavlov, 1928, p. 157). For example, a dog that has been conditioned to salivate to a bell of a certain tone will also salivate to bells of differing tones. The ability of the neighboring stimuli to produce the response varies with the degree of similarity to the original CS. Pavlov believed that we observe stimulus generalization because of an underlying physiological process he called *irradiation.* The initial stimulus excites a certain part of the brain which then irradiates, or spreads, over other regions of the cerebrum (1928, p. 157).

Discrimination. Initial generalization gradually gives way to a process of differentiation. For example, if one continues to ring bells of different tones (without presenting food), the dog begins to respond more selectively, restricting its responses to the tones that most closely resemble the original CS. One can also actively produce differentiation by pairing one tone with food while presenting another tone without food. This would be called an experiment in stimulus discrimination (Pavlov, 1927, pp. 118-30).

Higher order conditioning. Pavlov showed, finally, that once he had solidly conditioned a dog to a CS, he could then use the CS alone to establish a connection to yet another neutral stimulus. For example, in one experiment (1927, pp. 33-34), Pavlov's student trained a dog to salivate to a bell and then paired the bell alone with a black square. After a number of trials, the black square alone produced salivation. This is called second-order conditioning. Pavlov found that in some cases he could also establish third-order conditioning, but he could not go beyond this point (1927, p. 34).

Evaluation

In a sense, Pavlov's basic idea was not new. In the seventeenth century, Locke had proposed that knowledge is based on associations. However, Pavlov went beyond Locke and uncovered several principles of association through empirical experiments. He took the theory of learning out of the realm of pure speculation. Pavlov, as we shall see, did not discover everything there is to know about conditioning; in particular, his brand of conditioning seems restricted to a certain range of innate responses. Nevertheless, he was the first to put learning theory on a firm scientific footing.

WATSON

Biographical Introduction

The man most responsible for making Pavlovian principles a part of the psychological mainstream was John B. Watson (1878-1958). Watson was born on a farm near Greenville, South Carolina. In school, he said, "I was lazy, somewhat insubordinate, and so far as I know, I never made above a passing grade." (1936, p. 271). Nevertheless, he went to college at Furman University and graduate school at the University of Chicago, where he began doing psychological research with animals. After earning his doctorate degree, he took a position at Johns Hopkins University in Baltimore, where he did his most productive work.

In 1913 Watson made a great impact on psychology by issuing a manifesto, "Psychology as the Behaviorist Views It." In this article, Watson argued that the study of consciousness through introspection has no place in psychology as a science. Psychology should abandon "the terms consciousness, mental states, mind, content, introspectively verifiable, imagery and the like" (1913, p. 166). Instead, its goal should be "the prediction and control of behavior" (1913, p. 158). In particular, it should only study stimuli, responses, and the formation of habits. In this way psychology could become a science like the other natural sciences.

A year later he read the works of Pavlov and the Russians on conditioned reflexes and made Pavlovian conditioning the cornerstone of his thinking. Then, in 1918, Watson began research on young children, becoming the first major psychologist to apply principles of learning to the problems of development.

In 1929 Watson's academic career came to an abrupt end. He was divorced from his wife, an event which became so widely and sensationally publicized that Johns Hopkins fired him. Watson remarried (Rosalie Raynor, a primary co-worker) and entered the business world. In order to get a good sense of business, he worked for a while as a coffee salesman and a clerk at Macy's. He continued to write, but now for magazines such as *Cosmopolitan, Harper's,* and *McCall's.* In these articles, he primarily advanced his ideas on child development.

Basic Concepts

Environmentalism. Although Watson studied physiological responses, he was basically an environmentalist. In his book *Behaviorism* (1924), he made the famous proposal:

> Give me a dozen healthy infants, well-formed, and my own specified world to bring them up in and I'll guarantee to take any one at random and train him to become any type of specialist I might select—doctor, lawyer, artist, merchant, chief, and yes, even beggar-man and thief, regardless of his talents, penchants, tendencies, abilities, vocations, and race of his ancestors (p. 104).

In the next sentence, Watson added that "I am going beyond my facts, and I admit it, but so have the advocates of the contrary and they have been doing it for many thousands of years" (p. 104).

Study of emotions. One of Watson's major interests was the conditioning of emotions. He claimed that at birth there are only three unlearned emotional reactions—fear, rage, and love. Actually, all we observe are three different physical responses, but for the sake of simplicity we can call them emotions.

Fear, Watson said (1924, p. 152-54), is observed when infants suddenly jump or start, breathe rapidly, clutch their hands, close their eyes, fall, and cry. There are only two unconditioned stimuli that elicit fear. One is a sudden noise; the other is loss of support (as when the baby's head is dropped). Yet older children are afraid of all kinds of things—strange people, rats, dogs, the dark, and so on. Therefore it must be that the stimuli evoking most fear reactions are learned. For example, a little boy is afraid of snakes because he was frightened by a loud scream when he saw one. The snake became a conditioned stimulus.

Rage is initially an unlearned response to the restriction of body movement. For example, if we grab a two-year-old girl, preventing her from going where she wants, she begins to scream and stiffens her body. She lies down stiff as a rod in the middle of the street and yells until she becomes blue in the face (1924, p. 154). Although rage is initially a reaction to one situation—being forcibly held—it later is expressed in a variety of situations; children become angry when told to wash their faces, sit on the toilet, get undressed, take a bath, and so on. Such commands elicit rage because they have been associated with physical restraint in these situations. For example, the child becomes angry when told to get undressed because this order was initially associated with being forcibly held.

Love is initially a response which is automatically elicited by the stroking of the skin, tickling, gentle rocking, and patting. The baby responds by smiling, laughing, gurgling and cooing, and other responses that we call affectionate, good-natured, and kindly (1924, p. 155). Although Watson had no use for Freud, he noted that such responses "are especially easy to bring about by the stimulation of what, for lack of a better term, we may call the erogenous zones, such as the nipples, the lips, and the sex organs" (1924, p. 155).

Infants initially do not love specific people, but they are conditioned to do so. The mother's face frequently appears along with patting, rocking, and stroking, so it becomes a conditioned stimulus which alone elicits the good feelings toward her. Later, other people associated with the mother in some way also elicit the same responses. Thus, tender or positive feelings toward others are learned through second-order conditioning.

Actually, much of Watson's writing on the emotions was very speculative and was unsystematic speculation at that. He said that the three basic emotions became attached to a variety of stimuli and that "we get marked additions to the responses and other modifications of them" (1924, p. 165), but he said little about how these further developments occur. Where Watson did become specific was in his experimental work. His major experiment was on the conditioning of fear in an 11-month-old infant he called Albert B.

Conditioning fear in Little Albert. Watson and Raynor (Watson, 1924, pp. 159-64) wanted to see if they could condition Albert to fear a white rat. At the beginning of the experiment, Albert showed no such fear. Next, the experimenter on four occasions presented the rat and simultaneously pounded a bar behind Albert's head, producing a startle response. On the fifth trial, Albert was shown the rat alone, and he puckered his face, whimpered, and withdrew. He had been conditioned to fear the rat. For good measure, the experimenter combined the rat and the pounding twice more, and on the next trial, when the rat was again presented alone, Albert cried and tried to crawl away as rapidly as he could.

A few days later, Watson and Raynor tested for stimulus generalization. They found that whereas Albert played with many objects, he feared anything furry. He cried or fretted whenever he saw a rabbit, dog, fur coat, cotton wool, or a Santa Claus mask, even though he previously had not been afraid of these things. Thus Albert's fear had generalized to all furry objects.

Practical Applications

One of Watson's major practical innovations was a method for deconditioning fears. He was not able to decondition Albert of his new fears, because Albert was an orphan who was adopted and taken out of town before this could be attempted. But Watson advised one of his colleagues, Mary Cover Jones, on procedures for eliminating the fears of another little boy, a three-year-old called Peter.

Peter seemed active and healthy in every respect except for his fears. He was scared of white rats, rabbits, fur coats, feathers, cotton wool, frogs, fish, and mechanical toys. As Watson noted, "One might well think that Peter was merely Albert B. grown up, but Peter was a different child whose fears were 'home grown' " (1924, p. 173).

Jones tried a variety of methods, including having Peter watch other children play with a rabbit. But the procedure that she and Watson highlighted was the following. Peter was placed in his highchair and given a mid-afternoon snack. Then a caged white rabbit was displayed at a distance that did not disturb him. The next day, the rabbit was brought increasingly closer, until he showed a slight disturbance. That ended the day's treatment. The same thing was done day after day; the rabbit was brought closer and closer, with the experimenter taking care never to disturb Peter very much. Finally, Peter was able to eat with one hand while playing with the rabbit with the other. By similar means, Jones eliminated most of Peter's other fears as well.

Today, Jones's technique is known as a form of *behavior modification* called *systematic desensitization* (see Wolpe, 1969). The subject is relaxed and gradually introduced to the feared stimulus. The experimenter makes sure that the subject is at no time made to feel too anxious. Gradually, then, the subject learns to associate relaxed feelings, rather than fear, to the object or situation.

Watson did not confine his advice to therapeutic procedures for eliminating fears. He also had much to say on child-rearing, which he wanted to turn into a scientific enterprise. Watson recommended, among other things, that parents place babies on rigid schedules, and he insisted that they refrain from hugging, kissing, or caressing their babies. For when they do so, their children soon associate the very sight of the parent with indulgent responses and never learn to turn away from the parent and explore the world on their own (Watson, 1928, p. 81). Watson's advice was quite influential in the 1930s, but it was too extreme to last; under the influence of Spock, Bowlby, and others, parents relaxed their schedules somewhat and became more affectionate with their children. Nevertheless, Watson's more general goal—that of placing child-training on the firm foundation of scientific learning principles—remains a vital part of child care in the United States (Lomax *et al.*, 1978).

Evaluation

Watson convinced psychologists of Pavlov's relevance for human behavior. He inspired, for example, a number of investigations into the role of classical conditioning in infancy. It has proven more difficult to condition infants' reflexes than Watson suggested, but it generally does seem possible. And, as children grow, many of their emotional reactions to objects and people seem learned through classically conditioned associations (Liebert *et al.*, 1977, pp. 127-29).

At the same time, the usefulness of the classical conditioning paradigm seems limited. Classical conditioning seems to apply best to the conditioning of reflexes and innate responses (which may include many emotional reactions). It is questionable whether this kind of conditioning can also explain how we learn such active and complex skills as talking, using tools, dancing, or playing chess. When we master such skills, we are not limited to inborn reactions to stimuli, but we engage in a great deal of free, trial-and-error behavior, finding out what works best. Accordingly, learning theorists have developed other models of conditioning, the most influential of which is that of B. F. Skinner.

SKINNER AND OPERANT CONDITIONING

Biographical Introduction

B. F. Skinner was born in 1905 in the small town of Susquehanna, Pennsylvania. As a boy, he liked school and enjoyed building things, such as sleds, rafts, and wagons. He also wrote stories and poetry. After graduating from high school, he went to Hamilton College in New York. There he felt somewhat out of place, but he graduated Phi Beta Kappa with a major in English literature.

Skinner spent the next two years trying to become a writer, but he eventually decided that he could not succeed because he "had nothing important to say" (1967, p. 395). Because he was interested in human and animal behavior, he enrolled in the graduate psychology department at Harvard, where he began doing research and formulating his ideas on learning. Skinner has taught at the University of Minnesota (1936-1945), Indiana University (1945-1947), and Harvard, where he has been since 1947.

Despite his successful career as a scientist, Skinner never completely abandoned his earlier interests. For one thing, he continued to display his boyhood enthusiasm for building things. For example, when his first child was born, he decided to make a new, improved crib. This crib, which is sometimes called his "baby box," is a pleasantly heated place which does away with the necessity of excessive clothing and permits freer movement. It is not, as is sometimes thought, an apparatus for training babies. It is simply a more comfortable crib. Skinner's literary interests also re-emerged. In 1948, he published a novel, *Walden Two*, which describes a utopian community based on his principles of conditioning.

The Operant Model

Like Watson, Skinner is a strict behaviorist. He believes that psychology should dispense with any references to intangible mental states (such as goals, desires, or purposes); instead, it should confine itself to the study of overt behavior. Like Watson, in addition, Skinner is an environmentalist; although Skinner recognizes that organisms enter the world with genetic endowments, he is primarily concerned with how environments control behavior.

In contrast to Watson, however, Skinner's primary model of conditioning is not Pavlovian. The responses that Pavlov studied, Skinner says, are best thought of as *respondents*. These are responses that are automatically "elicited" by known stimuli. For example, the ingestion of food automatically elicits salivation, and a loud noise automatically elicits a startle response. Most respondents are probably simple reflexes.

A second class of behavior, which most interests Skinner, is called *operant*. In operant behavior, the animal is not harnessed in, like Pavlov's dogs, but moves freely about and "operates" on the environment. For example, in early experiments by Thorndike (1905), cats in a puzzle box would sniff, claw, and jump about until they hit upon the response—pulling a latch—which enabled them to get food. The successful response would then be more likely to recur. In such cases, we cannot always identify any prior stimulus that automatically elicits the responses. Rather, animals emit responses, some of which become more likely in the future because they have led to favorable *consequences*. Behavior, in Skinner's terms, is controlled by the reinforcing stimuli that follow it (Skinner, 1938, pp. 20-21; 1953, pp. 65-66. See also Munn *et al.*, 1974, pp. 208-10.) The two models, respondent and operant, are diagrammed below, in Figure 12.1.

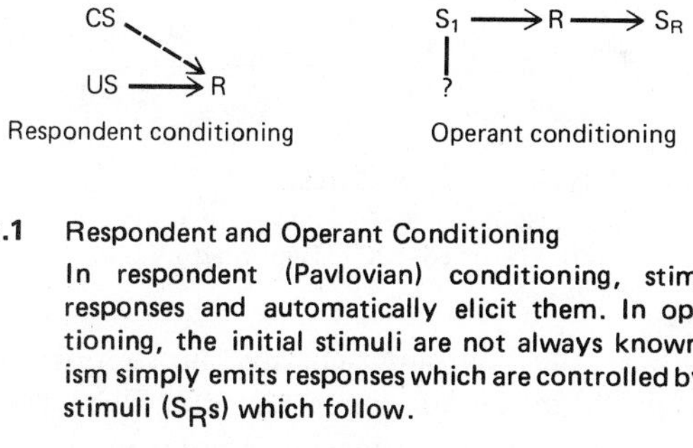

$$CS \searrow$$
$$US \longrightarrow R$$
Respondent conditioning

$$S_1 \longrightarrow R \longrightarrow S_R$$
$$\downarrow$$
$$?$$
Operant conditioning

FIGURE 12.1 Respondent and Operant Conditioning

In respondent (Pavlovian) conditioning, stimuli precede responses and automatically elicit them. In operant conditioning, the initial stimuli are not always known; the organism simply emits responses which are controlled by reinforcing stimuli (S_Rs) which follow.

To study operant conditioning, Skinner constructed an apparatus which is commonly referred to as a "Skinner box." This is a fairly small box in which an animal is free to roam about (see Figure 12.2). At one end there is a bar which, when pressed, automatically releases a pellet of food or water. The animal, such as a rat, at first pokes around until it eventually presses the bar, and then it gets the reward. As time goes on, it presses the bar more frequently. The most important measure of learning, for Skinner, is the *rate* of responding; when responses are reinforced, their rates of occurrence increase. In Skinner's apparatus, the bar-presses are automatically registered on a graph, so the experimenter need not be present much of the time. The data are presented as a learning curve, illustrated in Figure 12.3.

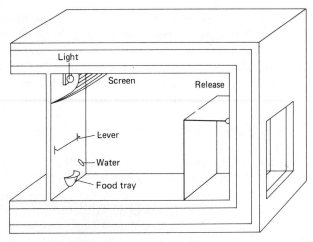

FIGURE 12.2   A Skinner Box. One side has been cut away to show the part occupied by the animal.

From Skinner, B. F., *The Behavior of Organisms,* p. 49. Copyright 1938, renewed 1966. Reprinted by permission of Prentice-Hall, Inc.

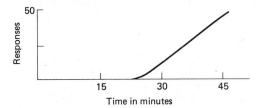

FIGURE 12.3   A Typical Learning Curve.

From Skinner, B. F. *The Behavior of Organisms,* Copyright 1938. Renewed 1966. Reprinted by permission of Prentice-Hall, Inc.

Skinner believes that operant behavior, in comparison to respondent behavior, plays a much greater role in human life. When we brush our teeth, drive a car, or read a book, our behavior is not automatically elicited by a specific stimulus. The mere sight of a book, for instance, does not elicit reading in the same way a bright light automatically elicits an eye-blink. We may or may not read the book, depending upon the consequences which have followed in the past. If reading books has brought us rewards, such as high grades, we are likely to engage in this behavior. Behavior is determined by its consequences (Munn *et al.*, 1974, p. 208).

Principles of Conditioning

Reinforcement and extinction. Skinnerians have performed numerous experiments showing that human behavior, beginning in infancy, can be controlled by reinforcing stimuli. For example, infants increase their rates of sucking when sucking results in sweet, as opposed to non-sweet liquid (Lipsitt, 1975). Similarly, infants' rates of smiling and vocalizations can be increased by making the behavior contingent upon rewards, such as the experimenter's smiles, caresses, and attention (Brackbill, 1958; Rheingold *et al.*, 1959).

In such experiments, one is dealing with different kinds of reinforcers. Some reinforcers, such as food or the removal of pain, are *primary reinforcers*; they have "natural" reinforcing properties. Other reinforcing stimuli, such as an adult's smiles, praise, or attention, are probably *conditioned reinforcers*; their effectiveness stems from their frequent association with primary reinforcers (Skinner, 1953, p. 78).

Operant behavior, like respondent behavior, also is subject to *extinction* (Skinner, 1953, p. 69). For example, because children do things "just to get attention" (1953, p. 78), one can extinguish undesirable behaviors, such as excessive crying or temper tantrums, by consistently withdrawing one's attention whenever they occur (Etzel and Gewirtz, 1967; Williams, 1959).

Operant behavior that has apparently been extinguished also may show *spontaneous recovery*. For example, a little boy whose temper tantrums had been extinguished through the withdrawal of attention began having tantrums once again when placed in a new situation (Williams, 1959). The behavior had to be extinguished further.

Immediacy of reinforcement. Skinner (1959, p. 133; 1953, p. 101) found that he could initially establish responses at the highest rates when he reinforced them promptly. For example, a rat will begin pressing a bar at a high rate only if it has promptly received a food pellet each time it has done so. As Bijou and Baer (1961, p. 44) point out, this principle has importance for child-rearing. If, for example, a father shows pleasure immediately after his son brings him his slippers, the boy is likely to repeat the behavior the next evening. If, however, the father is so engrossed in something else that he delays reinforcing his son's behavior for a few minutes, the boy's behavior will not be strengthened. In fact, what gets strengthened is the boy's behavior at the moment of reinforcement. If he is building blocks at that moment, it is block-building, not slipper-fetching, that gets reinforced.

Discriminative stimuli. We have said that operant conditioning may be described without any reference to initiating stimuli. This is true, but it does not mean that such stimuli are unimportant. Stimuli which precede responses may gain considerable control over them.

For example, Skinner (1953, pp. 107-8) reinforced a pigeon each time it stretched its neck. At this point, Skinner had no knowledge of any initial stimulus; he simply waited for the pigeon to emit the response and then reinforced it. Next, however, he reinforced the response only when a signal light was on. After a few trials, the pigeon stretched its neck much more frequently when the light was flashing than when it was off. The flashing light had become a *discriminative stimulus*. The light controlled the behavior because it set the occasion upon which the behavior was likely to be reinforced.

Skinner (1953, pp. 108-9) lists numerous examples to show how everyday behavior becomes attached to discriminative stimuli. In an orchard in which red apples are sweet and all others are sour, redness becomes a stimulus which sets the occasion upon which picking and eating will produce favorable outcomes. Similarly, we learn that a smile is an occasion upon which approaching another will meet with a positive response. When others frown, the same approach meets with aversive consequences, such as rebuffs. Insofar as this is true, the facial expressions of others become discriminative stimuli which control the likelihood that we will approach them.

Although discriminative stimuli do exert a considerable control, it must be emphasized that this control is not automatic, as in the case of respondent conditioning. In Pavlov's experiments, prior stimuli automatically elicit responses; in operant conditioning, such stimuli only make responses more *likely*.

Generalization. In operant conditioning, as in respondent conditioning, there is a process of *stimulus generalization* (Skinner, 1953, p. 132). Suppose a little girl has been reinforced for saying "Da da" at the sight of her father, but not when she is looking at her mother or siblings. The father has become a discriminative stimulus. It is not unusual, however, to find the girl saying "Da da" when she sees any man at all, such as strangers on the street. The stimulus has generalized. Her parents must now teach her to make a finer discrimination. They might say, "That's right," when she utters "Da da" in the presence of her father, but not when she looks at any other man.

Similarly, we can observe *response generalization*. It has been shown, for example, that when children are reinforced for using one part of speech, such as plurals, they begin uttering new plurals—even though they haven't received reinforcement for those particular words. Reinforcement influences not only particular responses but those of the same general class (Lovaas, 1977, pp. 112-13).

Shaping. Operant behavior is not acquired in all-or-nothing packages. It is usually learned gradually, little by little. Even teaching a pigeon to peck a spot on the wall, Skinner (1953, p. 92) shows, must be gradually shaped. If we place a pigeon in a box and wait for it to peck the spot, we may have to wait days or even weeks. Much of the time, the pigeon doesn't even approach the spot. So we must shape its behavior. First, we give the bird food when it turns in the direction of the spot. This increases the frequency of this behavior. Next, we withhold food until it makes a slight movement in the right direction. We then keep reinforcing positions closer and closer to the spot, until the bird is facing it. At this point, we can reinforce head movements, first giving food for any forward movement and finally reinforcing the bird only when it actually pecks the spot. Through this procedure, we gradually shape the desired response. Shaping is also called the "method of approximations," because reinforcement is made contingent upon better and better approximations of the desired response.

We probably teach many human skills in this bit-by-bit shaping process. For example, when we teach a boy to swing a baseball bat, we first say "Good" when he gets his hands into the right grip. We then say "Right" when he lifts his bat in the correct position over his shoulder. We then work on his stance, a level swing, and so on—gradually shaping the complete behavior.

Behavior chains. Although behavior may be shaped bit by bit, it also develops into longer, integrated response-chains. For example, batting in baseball involves picking up the bat, getting the right grip and stance, watching for the right pitch, swinging, running the bases, and so on. Skinnerians attempt to examine each step in terms of reinforcements and stimuli. Reaching for the bat is reinforced by obtaining it, which also serves as a stimulus for the next act, getting the right grip. Once the hands are placed on the bat, one gets a certain "feel" which one recognizes as the proper grip. This "feel" is a reinforcement, and it also signals the next action, pulling the bat over the shoulder. A little later, the sensation of the bat squarely striking the ball is a reinforcement for the swing, and it also signals the next action, running the bases. When a boy or girl has become a good hitter, the entire sequence is often performed in a smooth, integrated fashion (see Munn *et al.,* 1974, pp. 220-24, Reynolds, 1968, pp. 53-56, and Bijou, 1976 for accounts of chaining sequences).

Schedules of reinforcement. Skinner (1953, p. 99) observes that our everyday behavior is rarely reinforced *continuously,* every time; instead, it is reinforced *intermittently.* We do not find good snow everytime we go skiing or have fun every time we go to a party. Accordingly, Skinner has studied the effects of different schedules of intermittent reinforcement.

Intermittent reinforcement may be set up on a *fixed interval* schedule, such that the organism receives a reward for the first response after a specified period of time. For instance, a pigeon receives food after pecking a disc, but must wait three minutes before its next peck is rewarded, then three more minutes, and so on. The rate of responding on this schedule is generally low. Higher rates are produced by *fixed ratio* schedules, as when the pigeon gets food after every fifth peck. On both schedules, however, there is a lull in responding immediately after reinforcement. It is as if the organism knows it has a long way to go before the next reinforcement (Skinner, 1953, p. 103). Students often experience this effect immediately after completing a long term paper—it is difficult to get started on another assignment.

The lulls produced by fixed schedules can be avoided by varying reinforcement in unpredictable ways. On *variable interval* schedules, reinforcement is administered after an average length of time, but the intervals are mixed up. With *variable ratio* schedules, we vary the number of responses needed to produce a reward. When put on these two schedules, organisms consistently respond at high rates, especially on variable ratio schedules. They keep responding because a reward might come at any time.

One of Skinner's most important findings is that intermittently reinforced behavior, in comparison to that which is continuously reinforced, is much more difficult to extinguish. This is why many of our children's undesirable behaviors are so difficult to stop. For example, we might be able to resist a child's nagging or demanding behavior most of the time, but if we only yield every once in a while, the child will persist with it (Bijou and Baer, 1961, p. 62).

If we wish to begin teaching a desirable form of behavior, it is usually best to begin with continuous reinforcement; this is the most efficient way to get the behavior started. However, if we also wish to make the behavior last, we might at some point switch to an intermittent schedule (Bijou and Baer, 1961, p. 62).

Punishment. So far, we have been discussing ways to control behavior through positive reinforcements, such as food, attention, or praise. It also is possible to control behavior through *negative reinforcement* or through *punishment*. Technically, negative reinforcement and punishment are different. Negative reinforcement refers to the removal of an aversive stimulus (Skinner, 1953, p. 73). Basically, what is strengthened in this way is a tendency to escape, as when a girl standing on a diving board learns to escape the taunts of her peers by diving into the water (1953, p. 173).

When we punish, on the other hand, we do not strengthen behavior but try to eliminate it. Punishment, Skinner thinks, is "the commonest technique of control in modern life. The pattern is familiar: if a man does not behave as you wish, knock him down; if a child misbehaves, spank him; if the people of a country misbehave, bomb them" (1953, p. 182).

Punishment, however, does not always work. In an early experiment, Skinner (1938) found that when he punished rats for bar-pressing (by having the bar swing back and smack them on the legs), he only temporarily suppressed the response. In the long run, punishment did not eliminate the response any faster than did extinction. Other studies (e.g., Estes, 1944) have obtained similar results, and the findings conform to everyday experience. Parents who hit their children get them to behave for a while, but the parents find that the misconduct reappears later on.

Skinner also objects to punishment because it produces unwanted side effects. For example, a child who is scolded in school may soon appear inhibited and conflicted. The child seems torn between working and avoiding work because of the feared consequences. The boy or girl may start and stop, become distracted, and behave in other awkward ways (Skinner, 1953, pp. 190-91).

Some researchers believe that Skinner has overstated the case against punishment. In some instances, punishment will in fact completely eliminate responses. This is especially true when the punishment is extremely painful. Also, punishment can be effective when it is promptly administered, and when the organism can make alternative responses which are then rewarded (Liebert *et al.*, 1977, pp. 138-41). Nevertheless, the effects of punishment are often puzzling and undesirable.

Skinner recommends that instead of punishing children, we try extinction. "If the child's behavior is strong only because it has been reinforced by 'getting a rise out of' the parent, it will disappear when this consequence is no longer forthcoming" (1953, p. 192). Skinnerians often suggest that we combine extinction for undesirable behavior with positive reinforcement for desirable behavior. In one study, for example, teachers simply ignored nursery school children whenever they were aggressive and gave them praise and attention whenever they were peaceful or cooperative. The result was a quieter classroom (Brown and Elliot, 1965).

Internal Events: Thoughts, Feelings, and Drives

Thoughts. It is sometimes said that Skinner proposes an "empty organism" theory. He only examines overt responses and ignores internal states. This characterization is accurate but slightly oversimplified. Skinner does not deny that an inner world exists. We do have inner sensations, such as the pain from a toothache. We also can be said to think. Thinking is merely a weaker or more covert form of behavior. For example, we may talk to ourselves silently instead of out loud, or we may think out our moves silently in a chess game. However, such private events have no place in scientific psychology unless we can find ways of making them public and measuring them (Skinner, 1974, pp. 16-17, and Ch. 7).

Skinner is particularly distressed by our tendency to treat thoughts as the causes of behavior. We say that we went to the store because we "got an idea to do so" or that a pigeon pecked a disc because it "anticipated" food. However, we are in error when we speak in this way. We go to stores, and pigeons peck discs, only because these actions have led to past reinforcements. Any discussion of goals or expectations is superfluous. Worse, it diverts us from the true explanation of behavior—the controlling effect of the environment (Skinner, 1969, pp. 240-41; 1974, pp. 68-71).

Feelings. Skinner acknowledges that we have emotions, just as we have thoughts. However, feelings do not cause behavior any more than thoughts do. We might say we are going to the movies because "we want to" or because "we feel like it," but such statements explain nothing. If we go to the movies, it is because this behavior has been reinforced in the past (Skinner, 1971, p. 10).

Emotional responses themselves can be explained according to learning theory principles. In our discussion of Watson, we saw how emotional reactions might be learned through classical conditioning. Skinner believes that an operant analysis also is useful. Many emotions are the by-products of different reinforcement contingencies. Confidence, for example, is a by-product of frequent positive reinforcement. When we learn to hit a baseball sharply and consistently, we develop a sense of confidence and mastery (Skinner, 1974, p. 58). Conversely, we become depressed and lethargic when reinforcements are no longer forthcoming. On certain fixed ratio or interval schedules, we find it difficult to get going after receiving a reward because further rewards will not be coming for some time (Skinner, 1974, p. 59).

An operant analysis also helps us understand why various patterns of emotional behavior persist. If a little girl persistently behaves in an aggressive manner, it is important to know the consequences of this behavior. Do her actions succeed in getting attention or other children's toys? If so, her aggressiveness is likely to continue. Similarly, if displays of happiness, meakness, sympathy, fearfulness, and other emotional responses persist, it is because they have produced positive consequences (Skinner, 1969, pp. 129-30; Bijou and Baer, 1961, pp. 73-74).

Skinner believes, then, that we can understand emotions if we look at them as the products of environmental control. It is useless to consider emotions as intra-psychic causes of behavior, as the Freudians do. For example, a Freudian might talk about a man who fears sex because of anticipated punishment from an internal agency, the superego. To Skinner, such discussions get us nowhere. If we wish to understand why a person avoids sex, we must look at the past consequences of this behavior (Skinner, 1974, Ch. 10).

Drives. Skinner's refusal to look for causes of behavior within the organism has led to certain difficulties. In particular, he has had trouble with the concept of drive. Drives, such as hunger or thirst, would seem to refer to internal states which motivate behavior, and Skinner himself deprives his animals of food and water in order to make reinforcements effective.

Skinner argues that we do not need to conceive of drives as inner states, either mental or physiological. We simply specify the hours we deprive an animal of food or water and examine the effect of this operation on response rates (Skinner, 1953, p. 149).

Still, the drive concept has remained a thorn in the side of Skinnerians, and they have therefore searched for ways of conceptualizing reinforcement without reference to this concept. One interesting proposal has been made by Premack (e.g., 1961), who suggests that we think of reinforcement simply as the momentary probability of a response. Behavior that has a high probability of occurrence at the moment can serve as a reinforcer for behavior with a lower probability. If children are supposed to be eating their dinner but are busy playing instead, playing can be used as a reinforcer for eating. We simply say, "Eat some dinner and then play some more" (Homme and Totsi, 1965). Conceptualized in this way, eating and drinking have no special status as reinforcers. Eating and drinking, like any other actions, may or may not be good reinforcers, depending on their probabilities of occurrence at a particular time.

Species-Specific Behavior

Skinner argues, then, that we need not look inside the organism for the causes of behavior. Behavior is controlled by the external environment. There do seem to be, however, certain limitations to environmental control. In particular, each species has a particular genetic endowment which makes it easier to teach that species some things than other things. For example, it is hard to teach a rat to let go of objects, and it is hard to shape vocal behavior in non-human species (Skinner, 1969, p. 201).

Species-specific behavior, in Skinner's work, comes under the heading of *topography*. That is, the experimenter maps out a description of the behavior he or she can work with—for example, vocal behavior in humans. The topography is merely a description and does not constitute the most important part of the analysis, which is the way reinforcements shape and maintain behavior. Nevertheless, the topography is essential (Skinner, 1969, pp. 199-209).

In a larger sense, Skinner argues, even species-specific behavior is a product of environmental contingencies. For such behavior has become, in the course of evolution, part of the species' repertoire because it has helped that species survive in a certain environment. Thus, environments selectively reinforce all behavior—not only that in an animal's lifetime, but that in its species' evolutionary past (Skinner, 1969, pp. 199-209).

Practical Applications

Behavior modification. Skinner's research readily lends itself to practical applications, as we have seen in many of our illustrations. For example, we mentioned how Skinnerians might extinguish temper tantrums or get an unruly class to behave. The use of operant techniques to correct behavior problems is a branch of behavior modification. Operant techniques supplement the systematic desensitization procedures first employed by Watson and Jones.

Operant techniques have been applied to the treatment of autistic children, the kind we discussed in the chapter on Bettelheim. Autistic children, you will recall, are extremely isolated and engage in bizarre behaviors, such as self-stimulation. Many are mute, and others are echolalic (they merely echo what one says). Bettelheim thinks that they withdraw from others because they develop the feeling that they cannot influence others in any positive way. This interpretation can easily be framed in Skinnerian terms; the children lack positive reinforcement for their actions. However, the Skinnerian who has done the most work with these children, O. I. Lovaas (1973), is not at all certain about the cause of autism, and, more importantly, his treatment differs markedly from that of Bettelheim.

Bettelheim puts his faith in intrinsic growth forces and lets the child take the lead. He often permits socially inappropriate behavior as children gain autonomy and explore their problems. Lovaas, in contrast, takes charge. He realizes that children must act to learn, but he tries to control their actions. He wants to eliminate socially inappropriate behavior while reinforcing socially appropriate behavior (Lovaas *et al.*, 1973). If a child engages in echolalia or self-stimulation, Lovaas punishes the child or withdraws attention. When the child does something more appropriate, such as emitting correct speech, the child is rewarded. In the case of speech, the children often talk so little that their behavior must be very gradually shaped. At first they are rewarded for sounds that only roughly resemble words and are gradually reinforced for better and better approximations.

In general, the methods of Bettelheim and of Lovaas seem to be producing fairly equal outcomes. Lovaas's work does have the advantage of greater scientific precision, though. He tries to measure the results of each of his interventions.

Programmed instruction.   Skinner also has made a significant contribution to the education of normal children, through his invention of teaching machines and programmed instruction (Skinner, 1968). The teaching machine is a simple apparatus which permits one to read a brief passage, answer questions, and then, by turning a knob, see if one is correct. Actually, the machine itself is far less important than the programmed material it contains, and today the material is often presented in simple booklet form. To get an idea of how programmed instruction works, read the material below[2] and pretend to fill in the blanks. As you do so, cover the answers on the left side with a piece of paper, sliding it down just far enough to check your answers.

	1 Programmed instruction involves several basic principles of learning. One of these, called the principle of *small steps,* is based on the premise that new
small	information must be presented in _____ steps.
	2 The learner gradually acquires more and more
small steps	information, but always in _____ _____.
	3 Because active readers generally acquire more knowledge than passive readers, programmed instruction also is based on the principle of *active participation.* Writing key words as one is reading involves the
active	principle of _____ participation.
	4 While reading a book, an uninterested learner may slip into a passive state and discover that he cannot recall what he has just "read." In using programmed instruction the learner is prompted to remain alert by writing the key words, thus utilizing the princi-
active participation	ple of _____.

[2] From Munn, N. L., Fernald, L. D., and Fernald, P. S., *Introduction to Psychology*, 3rd ed., Boston: Houghton Mifflin Co., 1974, pp. 249-50. By permission of the publisher.

small steps	5 In these two techniques of programmed instruction, information is presented in _____ _____, and occasionally key words are missing, thus requiring
active participation	the learner's _____ _____ to complete the statements.
	6 A third principle, *immediate knowledge of results*, is illustrated when a professor returns quiz papers to his students at the end of the class in which they were written. These students receive almost imme-
knowledge	diate _____ of results.
	7 If a student makes an incorrect response at any point in programmed instruction, he discovers his mistake because the correct answer may be seen immediately after the frame, before the next one is considered. Thus, in programmed instruction, the
immediate; of results	learner receives _____knowledge _____ _____.
	8 Notice that in programmed instruction, unlike the evaluation of term papers, "immediate" does not mean a week or even a day but rather a few seconds. The reader of the program is continuously informed
immediate knowledge of results	concerning his progress; he receives _____ _____ _____ _____.
	9 Let us review the three techniques of programmed
small steps	instruction already considered. By means of _____
active participation	_____, the reader learns new material, which he acquires through _____ _____
immediate knowledge of results	followed by _____ _____ _____ _____.

Programmed instruction embodies several Skinnerian principles. First, it proceeds in small steps, because Skinner has found that the best way to establish new behavior is to shape it bit by bit. Second, the learner is active, because this is the natural condition of organisms. (Recall how Pavlov's dogs, in contrast, were harnessed in and simply reacted to stimuli.) Third, feedback is immediate because Skinner has found that learning is most rapid when promptly reinforced. (Reinforcement here is the knowledge that one's answer is correct.)

A sample of programmed reading for children is found in Figure 12.4 (p. 218).

In programmed instruction, students work independently and at their own pace. The instruction units are constructed so that each student may begin at a level that he or she can easily master. One does not want the student making many errors at first; for then the student will lack positive reinforcement for learning. As with shaping, one begins by reinforcing responses that are within the student's behavioral repertoire and gradually building up from there.

Surprisingly, programmed instruction bears some resemblance to Montessori methods. In both instruction is individualized, begins at the student's own level, and builds up skills gradually. Instruction is designed so that students can

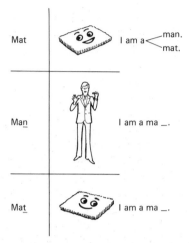

Mat	I am a < man. / mat.
Ma<u>n</u>	I am a ma _.
Ma<u>t</u>	I am a ma _.

FIGURE 12.4 Programmed instruction for children

Adapted from Sullivan, M. W., Programmed learning in reading. In A. D. Calvin (Ed.), *Programmed Instruction: Bold New Venture.* Bloomington, Ind.: Indiana University Press, 1969, p. 111. By permission of the publisher.

readily master each task. In both, the goal is not to tear down, through criticism or punishment, but to make learning a consistently positive experience.

However, programmed instruction also differs from Montessori methods. In the first place, programmed instruction involves materials which children read (see Figure 12.4), whereas Montessori materials are largely physical. Even when learning to read, Montessori children begin with sandpaper letters, metal insets, and so on. Montessori thought that young children find such physical activities more natural.

More fundamentally, there is a difference in the extent to which the child's work is free from adult direction. Montessori children choose their own tasks and work on them while the teacher steps into the background. It is hoped that they will discover for themselves how something is out of place, how cylinders fit, how water is poured, and whatever else that is important to them. In programmed instruction, in contrast, adult direction is pervasive. Although it might seem that children work independently on the booklets, in fact an adult (the program developer) has structured each small response. The child follows the adult's lead, repeatedly checking with this social authority to see if he or she is right. Children probably derive less sense that they are making their own discoveries about the world.

Nevertheless, it is important not to overlook the similarities between the two methods—especially the way both try to make learning a positive experience. One can even imagine Skinner approving of Montessori's physical tasks, albeit in his own terms. He would say that they work not because they allow for spontaneous discoveries, but because they allow children to make responses that readily result in positive feedback from the physical environment.

Evaluation

Skinner has considerably widened the scope of learning theory. After noting the limitations of classical conditioning, he has explored the nature of operant behavior, where the organism acts freely and is controlled by the consequences of its actions. In a brilliant series of studies, Skinner has shown how such control is exerted—by schedules of reinforcement, shaping, the influence of discriminative stimuli, and other factors. Furthermore, Skinner has amply demonstrated the practical importance of his ideas.

In the process, Skinner has stirred up controversies on many fronts. To some, his work lends itself to authoritarian practices; for he suggests ways to control, manipulate, and program others' behavior. Skinner's reply (e.g., 1974, p. 244) is that environments do, in fact, control behavior, and how we use our knowledge of this fact is up to us. We can create environments that suit humane purposes, or we can create ones that do not.

Developmentalists, too, often enter into heated, value-laden debates with Skinnerians. Developmentalists cringe at talk of controlling and changing children's behavior, when one should, instead, try to understand children and give them opportunities to grow on their own. To many Skinnerians, such sentiments are romantic and naive, for children chiefly develop through the molding influence of the external environment.

In a more objective vein, there are essentially three ways in which Skinner and writers in the developmental tradition disagree. First, developmental theorists often discuss *internal* events. Piaget describes complex mental structures, even though he might not be able to find direct evidence for all of them in any individual case. Freudians discuss internal events, such as unconscious fantasies, that we cannot directly observe at all. Skinner believes that such concepts divert us from scientific progress, which is made when we confine ourselves to the measurement of overt responses and environmental stimuli. On this point, however, even a number of learning theorists consider Skinner too extreme; they too believe that theories must include hypotheses regarding internal, cognitive events, even if we cannot directly measure them. In the next chapter, we will discuss a major example of a cognitive learning theory.

Second, developmental theorists and Skinnerians disagree on the meaning and importance of developmental *stages*—periods when children organize experience in very different ways. In Piaget's theory, for example, a child's stage is a crucial variable; it is a predictor of the kind of experience the child can learn from. A child at the sensori-motor level will not learn tasks that involve language, nor will a child beginning to master concrete operations learn much from lectures covering abstract theory.

Skinnerians doubt the validity of stages as general, distinct ways of thinking or behaving; for the environment shapes behavior in a gradual, continuous manner (Skinner, 1953, p. 91; Bijou, 1976, p. 2). Skinner does acknowledge that one must note the child's age in any experiment, just as one must note an animal's species and characteristic behavior (1969, p. 89). Age contributes to the "topography" of behavior; it helps describe the behavior that the experimenter sets about to shape or maintain. However, such information is still merely descriptive; it is secondary to environmental variables which control behavior. The question is whether the child's developmental status deserves this secondary role.[3]

A third issue dividing Skinner and developmental theorists is the most important of all. This issue concerns the *source* of behavioral change. Developmentalists contend that in crucial instances a child's thoughts, feelings, and actions develop spontaneously, from within. Behavior is not exclusively patterned by the external environment. Gesell, for example, believed that children stand, walk, talk, and so on from inner maturational promptings. Freudians, too, believe that development is governed by maturation, at least in part. For example, in Freud's view no one has to teach a little boy to become curious about sex. As his body matures, the phallic zone becomes very sensitive, arousing his interest in sexual matters. The child wants to compare his genitals to those of others and begins acting out adult sexual roles. The external environment may stimulate or suppress this behavior, but the initial interest comes from within.

Piaget is not a maturationist, but he also looks primarily to inner forces underlying developmental change. In his view, children's behavior is not structured by the environment but by children themselves. Children, out of a spontaneous interest in moderately novel events, construct increasingly complex and differentiated structures for dealing with the world.

Consider, for example, a baby girl who drops a block, hears the sound, and drops it again and again, making this new and interesting sound last. In Skinner's theory, the sound is a reinforcer which controls her behavior. However, this reinforcer will soon lose its effectiveness, for she will soon become interested in more complex outcomes (Kohlberg, 1969a). She may, for instance, begin listening for different sounds as she drops objects from different heights. For Piaget, one cannot look to external reinforcements as the determinants of behavior, for these often vary with the child's developing interests. For him, the main variable is the child's spontaneous curiosity about increasingly complex events.

Developmental theorists, then, try to conceptualize ways in which children grow and learn on their own, somewhat independent of others' teachings or

[3] For further discussions on the topic of stage and principles of conditioning, see S. White, 1965, and Gardner, 1978, p. 87.

external reinforcements. At the same time, no one can deny that environments also reinforce and control behavior to a considerable extent, and often in ways Skinner describes. Skinner's theory and research, moreover, have a clarity and elegant simplicity which others would do well to emulate. It is clear that Skinner's enormous contribution to scientific method and theory will be a lasting one.

13

Bandura's Social Learning Theory

BIOGRAPHICAL INTRODUCTION

The pioneering learning theorists usually developed their concepts by experimenting with animals in physical settings. They watched how animals ran through mazes, solved puzzle boxes, or learned to press levers in Skinner boxes. These situations were not social; there were no other animals present. Skinnerians and others then showed that the same principles also apply to learning in social contexts. Just as rats learn to press levers to get food, people learn to interact with others to obtain social rewards. However, some have questioned whether the analogy is really perfect. Bandura, in particular, argues that in social situations we learn a great deal through *imitation,* and that a full understanding of imitative learning requires several new concepts.[1]

Albert Bandura was born in 1925 in the province of Alberta, Canada. Like Skinner, he grew up in a very small town; his high school had only 20 students

[1] There are other social learning theories in addition to Bandura's. I have selected Bandura's because it is currently generating a great deal of research. The reader interested in other social learning approaches might begin with Baldwin (1967).

(Schultz, 1976, p. 302). Bandura attended the University of British Columbia as an undergraduate, and the University of Iowa as a graduate student in psychology. At Iowa he studied with Robert Sears, one of the pioneers in social learning theory. In 1953 Bandura joined the faculty at Stanford, where he has been ever since. Bandura has established a high reputation in psychology, and in 1974 he served as president of the American Psychological Association. His students consider him a modern generalist, a man whose knowledge spans across many fields in the social sciences (Gelfand, 1969, p. 186).

BASIC CONCEPTS

Observational Learning

In Skinner's theory, learning often appears to be a gradual process in which organisms must act to learn. Organisms emit responses, which are gradually shaped by their consequences. Bandura (1962), however, argues that in social situations people often learn much more rapidly simply by observing the behavior of others. When, for example, children learn new songs or play house just like their parents, they often reproduce long sequences of new behavior immediately. They appear to acquire large segments of new behavior all at once, through observation alone.

The power of observational learning is well-documented in the anthropological literature (Bandura and Walters, 1963, Ch. 2; Honigmann, 1967, p. 180). In one Guatemalan subculture, for example, girls learn to weave almost exclusively by watching models. The teacher demonstrates the operations of the textile machine, while the girl simply observes. Then, when the girl feels ready, she takes over, and she usually operates it skillfully on her very first try. She demonstrates, in Bandura's (1965a) term, "no-trial learning"; she acquires new behavior all at once, entirely through observation. She does not need to fumble through any tedious process of trial-and-error learning with differential reinforcement for each small response.

When new behavior is acquired through observation alone, the learning appears to be *cognitive*. When, for example, the Guatemalan girl watches her teacher and then imitates her perfectly without any practice, she must rely on some inner representation of the behavior which guides her own performance. Thus Bandura, unlike Skinner, believes that learning theory must include internal cognitive variables.

Observation also teaches us the probable consequences of new behavior; we notice what happens when others try it. Bandura calls this process *vicarious reinforcement*. Vicarious reinforcement also is a cognitive process; we formulate expectations about the outcomes of our own behavior without any direct action on our part.

We learn from models of many kinds—not only from live models but from *symbolic* models, such as those we see on television or read about in books. Another form of symbolic modeling is verbal instruction, as when an instructor describes for us the actions for driving a car. In this case, the teacher's verbal descriptions, together with a demonstration, usually teach us most of what we need to know. This is fortunate, for if we had to learn to drive exclusively from the consequences of our own actions, few of us would survive the process (Bandura, 1962, p. 214, 241).

Let us now look more closely at the observational learning process, which Bandura divides into four subprocesses.

The Four Components of Observational Learning

1. Attentional processes. First of all, we cannot imitate a model unless we pay attention to the model. Models often attract our attention because they are distinctive, or because they possess the trappings of success, prestige, power, and other winsome qualities (Bandura, 1971, p. 17). Television is particularly successful at presenting models with engaging characteristics and exerts a powerful influence on our lives (Bandura, 1977, p. 25).

Attention also is governed by the psychological characteristics of observers, such as their needs and interests, but less is known about such variables (1977, p. 25).

2. Retention processes. Since we frequently imitate models some time after we have observed them, we must have some way of remembering their actions in symbolic form. Bandura (1965a; 1971, p. 17) thinks of symbolic processes in terms of *stimulus contiguity,* associations among stimuli that occur together. Suppose, for example, we watch a man use a new tool, a drill. He shows us how to fasten the bit, plug it in, and so on. Later, the sight of the drill alone arouses many associated images, and these guide our actions.

In the above example, the stimuli are all visual. However, we usually remember events, Bandura (1971, p. 18) says, by associating them with verbal codes. When, for example, we watch a motorist take a new route, we connect the route with words (e.g., "Route 1, then Exit 12 . . ."). Later, when we try to drive the route ourselves, the verbal codes help us follow it.

Young children, under the age of five years or so, are not yet accustomed to thinking in words, and probably must rely quite heavily on visual images. This limits their ability to imitate. One therefore can improve on their imitations by directing them to use verbal codes—that is, by asking them to give verbal descriptions of a model's behavior while they are watching it (Bandura, 1971, p. 19; Coates and Hartup, 1969). It is not clear, though, whether such instructions also help older children, who may engage in silent verbal coding spontaneously.

Verbal coding is distinguished from verbal rehearsal; rehearsal involves describing behavior over and over so one can remember it better on a later occasion. Rehearsal techniques seem to enhance recall for most everyone (Bandura, 1971, p. 21).

3. Motor reproduction processes. To reproduce behavior accurately, the person must have the necessary motor skills. For example, a boy might watch his father use a saw but find that he cannot imitate very well because he lacks the physical strength and agility. From observation alone, he picks up a new *pattern* of responses (e.g., how to set up the wood and where to place the saw), but no new physical abilities (e.g., cutting with power). The latter come only with physical growth and practice (Bandura, 1977, p. 27).

4. Reinforcement and motivational processes. Bandura, like cognitive learning theorists before him (Tolman, 1948), distinguishes between the *acquisition* and the *performance* of new responses. One can observe a model, and thereby acquire new knowledge, but one may or may not perform the responses. For example, a boy might hear his neighbor use some profane language, and thereby learn some new words, but the boy might not reproduce them himself.

Performances are governed by reinforcement and motivational variables; we will actually imitate another if we are likely to gain a reward. In part, it is our past history of *direct reinforcements* that matters. If, in the above example, the boy has himself received respect and admiration for swearing, he is likely to imitate his neighbor. If, on the other hand, he has been punished for swearing, he probably will hesitate to imitate him.

Performances also are influenced by *vicarious reinforcements*, the consequences one sees accrue to the model. If the boy sees his neighbor admired for swearing, the boy is likely to imitate him. If he sees the neighbor punished, he is less likely to do so (Bandura, 1971, p. 46; 1977, pp. 117-24).

Performances, finally, are partly governed by *self-reinforcements*, the evaluations we make of our own behavior. We will discuss this process in a later section.

Conclusion. To successfully imitate a model, then, we must 1) attend to the model, 2) have some way of retaining what we have seen in symbolic form, and 3) have the necessary motor skills to reproduce the behavior. If these conditions are met, we probably know how to imitate the model. Still, we might not do so. Our actual performances are governed by 4) reinforcement contingencies, many of which are of a vicarious sort.

In reality, these four components are not totally separate. Reinforcement processes, in particular, influence what we attend to. For example, we often attend to powerful, competent, prestigious models because we have found that imitating them, rather than inferior models, leads to more positive consequences.

SOCIALIZATION STUDIES

Bandura's four-part model gives a fine-grained analysis of imitative learning. On a broader level, Bandura's primary, if sometimes implicit, concern is the socialization process—the process by which societies induce their members to behave in socially acceptable ways.

Socialization is an inclusive process which influences almost every kind of behavior, even technical skills. Many American teenage boys, for example, feel that they will not fit into their social group unless they learn to drive a car. Automobile driving, however, is not something that is required by all cultures, and there are classes of social behavior that have broader relevance. For example, all cultures seem to try to teach their members when it is acceptable to express aggression. It also is likely that all cultures try to teach people certain modes of cooperation, sharing, and helping. Aggression and cooperative behavior, then, are "targets" of socialization in all cultures (Hetherington and Parke, 1977, p. 231). In the next few sections, we will sample social learning analyses of some of the target behaviors in the socialization process.

Aggression

Bandura (1967; Bandura and Walters, 1963) believes that the socialization of aggression, as well as other behavior, is partly a matter of operant conditioning. For example, parents and other socializing agents reward children when they express aggression in socially appropriate ways (e.g., in games or in hunting) and punish children when they express aggression in socially unacceptable ways (e.g., hitting younger children). But socializing agents also teach children a great deal by the kinds of models they present. Children observe aggressive models, notice when they are reinforced, and imitate accordingly. Bandura has examined this process in several experiments, one of which is now considered a classic.

In this study (Bandura, 1965b), four-year-olds individually watched a film in which a male model engaged in some moderately novel aggressive behavior. The model laid a Bobo doll on its side, sat on it, and punched it, shouting such things as, "Pow, right in the nose," and "Sockeroo . . . stay down" (pp. 590-91). Each child was assigned to one of three conditions, which meant that each child saw the same film but with different endings.

In the *aggression-rewarded* condition, the model was praised and given treats at the end of the film. A second adult called him a "strong champion" and gave him chocolate bars, soda pop, and the like (p. 591).

In the *aggression-punished* condition, the model was called a "big bully," swatted, and forced to cower away at the end of the film (p. 591).

In the third, *no-consequences* condition, the model received neither rewards nor punishments for his aggressive behavior.

Immediately after the film, each child was escorted into a room with a Bobo doll and other toys. The experimenters observed the child through a one-way mirror to see how often he or she would imitate the aggressive model.

The results indicated that those who had seen the model punished exhibited significantly fewer imitations than did those in the other two groups. Thus, vicarious punishment reduced the imitation of aggressive responses. There was no difference between the aggression-rewarded and no-consequences groups. This is often the finding with respect to behavior, such as aggression, that is typically prohibited. The observation that "nothing bad happens this time" prompts imitation just as readily as does vicarious reward (Bandura, 1969, p. 239).

The experiment also had a second, equally important phase. An experimenter came back into the room and told each child that he or she would get juice and a pretty sticker-picture for each additional response he or she could reproduce. This incentive completely eliminated the differences among the three groups. Now all the children—including those who had seen the model punished—imitated him to the same extent. Thus, vicarious punishment had only blocked the *performance* of new responses, not their *acquisition*. The children in the aggression-punished condition had learned new responses, but had not felt that it was wise to actually reproduce them until a new incentive was introduced.

One of Bandura's followers, Robert Liebert (Liebert *et al.*, 1977, p. 145), suggests that this experiment has implications for aggression in television and movies. Children are frequently exposed to actors who demonstrate clever ways of committing homicides and other crimes. The widespread showing of such films is justified by the fact that the criminals are usually caught in the end. However, Bandura's work suggests that children probably learn about criminal behavior nonetheless, and only inhibit such behavior until a time when environmental contingencies clearly favor their occurrence.

In the above experiment, children performed *newly* acquired responses. Models also can influence the performance of *previously learned* behavior of the same general class. For example, a boy might watch a violent movie and then act roughly toward his sister. He does not actually imitate the behavior he saw in the film, but he feels freer to engage in previously learned behavior of the same kind. In such cases, we say that the behavior has been *disinhibited*. Models also may *inhibit* previously learned behavior, as when a girl sees a boy punished in class and therefore decides to check her impulse to do something else of a mischievous nature (Bandura and Walters, 1963, p. 72; Liebert *et al.*, 1977, pp. 146-47).

Sex Roles

During socialization, children are taught to behave in sex-appropriate ways. Societies encourage boys to develop "masculine" traits and girls to develop "feminine" traits.

It is possible, of course, that sex traits are also, in part, genetically linked. Social learning theorists do not deny this possibility. However, they believe that more is to be gained from the study of socialization processes and the role of imitation in particular (Bandura and Walters, 1963, pp. 26-29; Mischel, 1970).

In the learning of sex roles, the acquisition/performance distinction is especially important (Mischel, 1970). Children frequently learn, through observation, the behavior of both sexes; however, they usually perform only the behavior appropriate to their own sex because this is what they have been reinforced to do. For example, Margaret Mead (1964) tells how Eskimo boys are encouraged to practice hunting and building snow houses, whereas girls are not. So ordinarily only the boys engage in these activities. Nevertheless, the girls watch the boys, and in emergencies they can execute the same skills, though not as well. They pick up the skills through observation alone, although they are not as adept as the boys because they have not practiced them.

Social reinforcements, then, only restrict the range of skills that boys and girls practice—not what they observe. After a while, however, children may even stop making quite as careful observations of opposite-sex models (Maccoby and Wilson, 1957). Perhaps it is as if a boy, for instance, were to say: "I seem to make people happy only when I try things boys do; so I'm only going to pay close attention to the things boys do." Thus, social reinforcements may even influence the observational process itself. Support for this possibility comes from an experiment which showed that when children earned rewards for imitating a particular model, they had better recall for that model's subsequent behavior (Grusec and Brinker, 1972).[2]

Self-Reinforcement

As people become socialized, they depend less on external rewards and punishments and establish patterns of self-regulation. That is, they establish their own internal standards and reward and punish themselves in accordance with them. For example, a woman might criticize herself for a moral transgression that no one else is even aware of. She punishes herself because her behavior violated her own standards.

Bandura has been most interested in how people reinforce their own performances as they strive for success and achievement. Some people set exceedingly high achievement goals and reward themselves only when they meet

[2] For a good review of the literature on sex differences, as well as various theoretical interpretations, see Maccoby and Jacklin, 1974.

them. An artist, for example, might approve of his own work only after he has corrected flaws that others would never detect. Others are satisfied with less perfect work.

How are self-evaluative standards acquired? In part, Bandura believes, they are the product of direct rewards and punishments. For example, parents might give their daughter approval only when she earns very high grades, and after a while she adopts this high standard as her own.

However, Bandura's focus, once again, is on the influence of models, which he first demonstrated in a key experiment (Bandura and Kupers, 1964). Basically, the design was as follows. Seven- to nine-year-old children watched a model play a bowling game. In one condition, the model rewarded himself with candy only when he obtained a high score, saying things such as, "I deserve some M&Ms for that high score" (1964, p. 3). Otherwise, he made self-critical remarks, such as, "That does not deserve an M&M treat" (1964, p. 4). In a second condition, the model followed a lower criterion for self-reward. There was also a control group of children who saw no model. The main finding was that afterward, when the children played the game alone, they adopted the patterns of self-reward of the model to whom they had been exposed. The children in the control group, who had seen no model, adopted no consistent standards; they took treats whenever they felt like it. The models, then, exerted clear-cut effects on the children's self-evaluative behavior. Subsequent research has complicated the picture somewhat, but it has not altered the general finding (Bandura, 1977, pp. 134-36).

Prosocial Behavior

In the last decade, there has been considerable interest in the nature and roots of prosocial behavior—acts of sharing, helping, cooperation, and altruism. Social learning theorists have taken the lead in this area, showing that prosocial behavior can be readily influenced by exposure to the appropriate models. In a typical study (Rushton, 1975), seven- to 11-year-old children watched an adult model play a bowling game and donate some of his winnings to a "needy children's fund." Immediately afterward, these children played the game alone, and they themselves made many donations—far more than did a control group who had not seen the altruistic model. Furthermore, the children who had observed the model still donated more two months later, even when placed in a different room with a different experimenter. Evidently, even a relatively brief exposure to a generous model exerts a fairly permanent effect on children's sharing.

Numerous other experiments have shown that models influence not only children's sharing but their helpfulness toward others in distress, their cooperativeness, and their concern for the feelings of others (Mussen and Eisenberg-Berg, 1977, pp. 79-90; Bryan, 1975). The experimental findings in this area also seem supported by more naturalistic studies, in which parental behavior is linked to their children's altruism (Mussen and Eisenberg-Berg, 1977, pp. 86-90).

Practicing and preaching. Socializing agents teach children not only by behavioral example, but by preaching virtue and telling children how to behave. Such verbal techniques have been most fully explored in research on prosocial behavior, so a brief review might be in order.[3]

First of all, preaching seems ineffective unless it is forceful. If an adult simply says, "It is nice to share," the child will be far more influenced by what the adult actually does. If the adult shares, so will the child—regardless of whether the adult preaches altruism or greed (Bryan and Walbeck, 1970). When, however, the preaching becomes stronger, taking the form of long emotional sermons and commands, it can be effective (Mussen and Eisenberg-Berg, 1977, pp. 151-52).

Commands, however, are coercive and may backfire, as found in a study by G. M. White (1972). In this experiment some children took turns bowling with an adult who *told* them to share some of their winnings with needy children. Other children were simply given the opportunity to follow an altruistic example. The immediate result was that the children who were ordered to share did so to a greater extent, even when playing alone. In a posttest, however, these children's sharing decreased sharply, and they displayed a greater incidence of stealing, perhaps reflecting their resentment against the coercive technique.

Conclusion

We have now seen how models can influence four kinds of behavior—aggression, sex roles, patterns of self-reinforcement, and prosocial behavior. Our review only samples the research on these topics, but it does indicate the powerful effects that models can have. As Bandura (1977, p. 88) has said, "One can get people . . . to engage in most any course of action by having such conduct exemplified." Models also can alter, for example, the extent to which children delay gratification, resist temptation, and use certain types of moral thinking— all important aspects of the socialization process. We cannot review all these topics, but we will discuss moral judgment in the following section.

ABSTRACT MODELING AND PIAGET'S STAGES

It is commonly assumed that imitation can only produce exact copies of a model's behavior. If this were so, the influence of models would be quite limited. However, observers also induce the rules underlying a model's behavior

---

[3] For studies on the transmission of achievement standards through verbal techniques, see Bandura, 1977, p. 137.

and use the rules to create behavior which, while similar to that of the model, is actually new. For example, children induce the grammatical rule for forming plurals (add the "s" sound), and they use it to generate new sentences (Bandura, 1971, pp. 33-36).

Through abstract modeling, children can learn many kinds of concepts and rules, including those Piaget has investigated. Bandura's view of this process, however, differs from that of Piaget. Like Piaget, he sees the child as an active cognitive agent; the child induces rules and grasps concepts. However, Bandura's emphasis is much more on the way the external environment—particularly models—influence the kinds of concepts children formulate. Let us compare the two theories a bit more thoroughly.

Piaget, you will recall, sees development as an inner process which unfolds somewhat independently from external teachings. Children learn much on their own, out of an intrinsic interest in the world. As they do so, their thinking undergoes a series of internal transformations, called stages. These stages, in turn, determine what children will learn; children most eagerly and readily grasp information that is just beyond their current stage. This holds for imitation too; children are spontaneously interested in models whose behavior is moderately novel or slightly more complex than their own (Kohlberg, 1966b; Kuhn, 1974). This is why we often see a young child enthusiastically tagging after a somewhat older one, trying to do the same things. Thus, Piagetians do not spend much time examining the modeling influences in a child's life, but study, instead, children's thinking at each stage, for this determines the models they will select and benefit from.

Bandura, in contrast, is primarily an environmentalist. In his view, development is not a process of inner growth and spontaneous discovery. Rather, the child's mind is structured by its exposure to social models and by social training practices (Bandura, 1977, p. 183; Bandura and Walters, 1963, p. 44).

As an environmentalist, he sees two basic things wrong with Piaget's theory. First, Bandura doubts whether children learn much out of an intrinsic interest in the world. Instead, they must be motivated by extrinsic inducement, such as a teacher's praise or criticism. After a while children begin applying praise and criticism to their own work, but these evaluations are essentially internalizations of external standards (Bandura, 1977, pp. 106, 164-65).

Second, Bandura strenuously objects to Piaget's assertion that internal structures, or stages, determine what children will learn. It is true, Bandura acknowledges, that we cannot teach children lessons that exceed their understanding. However, we do not need to consult Piaget's elaborate stage theory, with its assumption of fixed sequences, to find out what children will imitate. In fact, it is not so much that internal cognitive structures determine what children will imitate, but the other way around: Children's cognitive structures are themselves determined by the models they are exposed to (1977, pp. 43-47).

In support of their position, social learning theorists have tried to show that Piaget's stages can be readily modified by social learning procedures. One study (Bandura and McDonald, 1963) dealt with Piaget's stages of moral reasoning.

Moral reasoning. Piaget, you will recall, proposed a two-stage theory of moral judgment, one aspect of which concerns consequences versus intentions. That is, younger children tend to judge wrongdoing in terms of its consequences, whereas older children base their judgments on the intentions behind the act. For example, a young child is likely to say that a boy who broke 15 cups by accident is naughtier than one who was trying to steal but only broke one cup. The young child focuses on the consequences—the amount of damage. The older child, in contrast, usually puts more weight on the underlying motive.

Bandura gave five- to 11-year-old children 12 such items, and found the age shift that Piaget and others (Kohlberg, 1969a) have documented. However, Bandura emphasized, children of all ages evidenced at least some reasoning of both kinds, suggesting that the stages are not rigidly demarcated.

Following this pretest, Bandura tried to show that the children's thinking could be altered by modeling influences. In a key part of the experiment, children individually observed an adult model who was praised for giving responses *contrary* to their own dominant mode. If, for example, a child had typically judged wrongdoing in terms of intentions, the model always based her judgment on the consequences. An experimenter presented the model with a moral dilemma, praised her when she gave her answer, and gave the child a turn on a new item. Taking turns in this way, the model and child each responded to 12 new items (different from the pretest items).

This training procedure did have a strong effect. Prior to the training, children gave one type of moral response only about 20 percent of the time; during the treatment, this number increased to an average of about 50 percent.

The experiment also included an immediate posttest in which the children responded once again to the pretest items. The results indicated that the children persisted with their new mode of responding (about 38 to 53 percent of the time).

The study, Bandura says, shows that "the so-called developmental stages were readily altered by the provision of adult models . . ." (Bandura and Walters, 1963, p. 209). There seems to be nothing fixed or invariant about them.

A group of cognitive-developmentalists (Cowan *et al.*, 1969) replicated Bandura's modeling treatment and obtained similar results. Nevertheless, cognitive-developmentalists have viewed the study suspiciously. They acknowledge that modeling can influence cognitive stages, but the influence should be small. This is because stages represent broad, deeply rooted cognitive structures. We cannot, in theory, effortlessly get a child to reason in any way we wish. And when we do produce change, it should be primarily in the direction of the stage sequence—one stage forward. Several experiments have, in fact, found that these are the kinds of changes that do occur when Kohlberg's, rather than Piaget's, stages are used (Gardner, 1978, p. 208). This, Kohlberg (1969a) argues, is because his stages represent broader cognitive structures than Piaget's moral stages do, so his are harder to change. However, as Bandura (1977, p. 45) notes, the modeling influences in these studies were brief and weak.

We can see, then, that Bandura's experiment has stirred up a good deal of controversy. He has presented a serious challenge to cognitive stage theory.

Conservation. Social learning theorists also have tried to show that conservation can be altered through modeling. Some of the best experiments have been performed by Rosenthal and Zimmerman (Rosenthal and Zimmerman, 1972; Zimmerman and Rosenthal, 1974). In a typical study (1972), five- and six-year-olds were first given a battery of conservation tasks (e.g., conservation of number, liquid, and weight). Those who showed little knowledge of conservation were assigned to one of two groups. Some children were assigned to a modeling group in which they observed an adult model perform the tasks correctly. Others were placed in a control condition; they saw no model. All the children were then readministered the same test battery and were also given a new series of items to test for generalization. Actually, the generalization tasks were of the same kind (e.g., conservation of number, liquid, and weight); they differed only in colors and shapes and in other superficial ways.

The results indicated that the modeling group outperformed the control group; they gave significantly more correct conservation responses. This was true for both the repeated tasks and the generalization tasks.

In this study, the amount of change produced by the modeling treatment was impressive, but this is not always the case. In another study (Rosenthal and Zimmerman, 1972), for example, four-year-olds acquired new conservation skills only to a very modest degree.

On the basis of their several experiments, Rosenthal and Zimmerman enthusiastically conclude that modeling can produce rapid and substantial changes in conservation behavior (1972, p. 399; 1974, p. 268). Conservation skills, they believe, are probably the product of socialization—of the teachings of adults in the child's culture. Developmentalists would be more skeptical and would point to the weaker results with children who were not ready to benefit from the modeling experience.

FREUDIAN CONCEPTS OF IDENTIFICATION

Freudians, like social learning theorists, are quite interested in how we take on the characteristics of parents and other models. Freudians usually distinguish between imitation, the copying of isolated actions, and identification, a process by which we incorporate others' attributes in a more wholesale manner (Bronfrenbrenner, 1960).

Basically, Freud wrote about two kinds of identification. One kind is *anaclitic identification* (Freud, 1914; 1917; 1923, p. 19). "Anaclitic" means "leans on" and refers to a situation in which we are dependent on others for love and care. Freud suggested that we identify with those whom we love, especially when they are absent, as an unconscious way of keeping them with us. We saw an example of this in Erikson's case of Peter. When Peter lost his nurse, he imagined himself pregnant like she was—this was his way of holding on to her.

Anaclitic identification is not restricted to situations of traumatic loss. Loved ones are almost always absent a certain amount, so we tend to identify with them as a way of keeping them.

Anaclitic identification begins early in life, in the oral stage. A second kind of identification, which might be called *rivalrous,* first emerges in dramatic form during the oedipal crisis. The little boy, in particular, becomes fearful of the consequences of rivaling his father and identifies with him instead. This is the boy's way of relieving his anxiety. Since it is often the father whom the boy fears, this identification is sometimes called "identification with the aggressor" (A. Freud, 1936).[4]

Bandura (1969) argues that the Freudian approach impedes any attempt to formulate a general theory of modeling. The Freudians introduce various distinctions—between identification and imitation, and between different kinds of identification, which have us searching for separate modeling processes at the outset. We do not get a chance to find out if more unitary processes are at work. Further, the identification processes, which are presumably so powerful, apply to only a very narrow range of situations. To identify with a model, we must be involved in an intimate relationship with the person. We must either be in need of the model's love or in an intense rivalry with the person. Yet we also emulate television personalities, sports heroes, and a host of other people whom we have never even met.

Bandura says that if we wish to formulate a general theory, we should rid ourselves of all the Freudian distinctions. We should consider identification and imitation as a single process—that of matching the behavior of models—and see if we can explain this process in terms of the same motivational variables. In Bandura's view, we emulate all models, even those we do not know, to maximize positive reinforcements (e.g., to gain the rewards that they seem to possess).

There is yet another facet of the Freudian theory which complicates any simple, general theory of modeling: Identification is partly tied to maturational stages. Let us look briefly at this stage-dependent view of modeling and then discuss Bandura's critique.

---

[4] For reviews of the research on Freudian concepts of identification, some of which has been conducted by Bandura himself, see Bandura, 1969, and Hoffman, 1970.

At the *oral* stage, when the baby is biologically dependent on caretakers, anaclitic identification predominates. Infants begin to "take in" the parent to keep the parent when absent. Anaclitic identifications persist throughout life, but in the later stages modeling also shifts somewhat.

At the *anal* stage, in which maturation ushers in a strong need for autonomy and even negativism, children often display a distinctive reluctance to imitate. Tell a two-year-old, "Do it this way," and the likely response will be "No!" (Kohlberg, 1969a, pp. 432-33).

At the *phallic* or *oedipal* stage, rivalrous identification comes into play and exerts a great influence on character formation.

During the *latency* stage, when Erikson says children are developing a sense of industry, children are most interested in models that exemplify teachable and realistic skills (Kohlberg, 1969a, p. 451).

During *adolescence,* teenagers often reject certain models, especially parental models, as they try to establish independence from them.

We see, then, that modeling varies from stage to stage. So if one wants to know what modeling is like, one must examine the stage in question—a task which would certainly complicate Bandura's environmentalistic theory. Not surprisingly, then, Bandura (1964) argues that the Freudian stages have no more validity than Piaget's. He focuses his critique on the Freudian view of adolescence as a separate stage of life.

According to Sigmund and Anna Freud, adolescence is a very stressful period. The maturation of powerful sexual and aggressive drives throws the young person into a state of turmoil. These drives revive old oedipal conflicts, and the teenager thinks that the only solution is to gain independence from the immediate family. As a way of gaining distance, young people often rebel against their parents. They reject parental models and turn to the peer group for support and guidance.

According to Bandura (1964), this view is largely a myth. He cites evidence that suggests that for most North American youth, adolescence is no more stressful than childhood. Nor is it a period of sudden independence-seeking; for parents already had taught their sons and daughters to become increasingly independent during childhood. Nor, finally, is it a period of rebellion against parents; for most teenagers continue to accept the advice and values of their parents, and they join peer groups that uphold and reinforce these values.

For most North American youth, then, adolescence represents a continuity with the past—not a separate developmental stage. In some other cultures, it is true, adolescence is more distinctive; it may be characterized, for example, by special initiation rites and rituals. However, these are cultural phenomena and are not universal.

One might agree with Bandura's argument and still wonder if there isn't one major maturational development which makes adolescence distinctive—the onset of puberty. Surely the adolescent's life is now different if only because he or she now has sex constantly on the mind. Bandura argues, however, that even

sex is basically a cultural matter, and cultures determine whether there is anything new or different about sexuality in adolescence. For most American
youth, the culture does expect a rather sudden interest in dating and the
opposite sex. However, this is not always so. Not long ago, for example,
Samoans were subtly encouraged to engage in a good deal of sex play during
childhood, so their heterosexual experiments at puberty represented nothing
new (Mead, 1928).

Adolescence, then, does not qualify as a universal stage of life. It differs
from other periods only when cultures make it different, and this is by no means
always the case. If we wish to understand adolescent behavior—or behavior
during any other phase of development—we need to examine cultural and social-
learning variables. We do not need to consider maturational stages.

PRACTICAL IMPLICATIONS

Bandura's work should do a good deal to increase our awareness of the
importance of models in child-rearing and education. Although most parents and
teachers already are somewhat aware of the fact that they teach by example,
they probably have also overlooked just how influential modeling can be. A case
in point is physical punishment. Many parents try to prevent their children from
fighting by spanking them when they fight—only to find, it seems, that their
children fight all the more (Bandura and Walters, 1963, p. 129). The likely
explanation is that the parents, by spanking, are inadvertently providing a good
demonstration of how to hurt others (Bandura, 1967). Similarly, whenever we
find that we are unable to rid a child of some distressing bit of behavior, we
might ask whether we have been inadvertently modeling the behavior ourselves.

Modeling, according to Bandura, takes many forms. The familiar kind is
behavioral modeling; we exemplify an activity by performing it. Modeling also
may be done verbally, as when we give instructions or issue commands. Social
learning researchers are evaluating the effectiveness of the various kinds of
modeling, and their findings should be of importance to parents and educators.
Of particular interest are studies such as White's (1972), which examined the
effects of commanding children to share. At first, the commands seemed to
work, but their impact diminished over time, and the commands also produced
resentment and rebelliousness. In the long run, we may do better simply to
model generosity and helpfulness through our own behavior. Then children can
follow our example without feeling forced to do so.

Social learning theorists also have shown that behavior is influenced not
only by personal or live models but by those presented in the mass media.
Filmed models, in particular, seem to exert a powerful impact, and one major
implication is that television, which many children watch for hours on end,

is shaping young lives. Social learning theorists have been especially concerned with the effects of televised violence on children and have found that, in fact, it does increase children's aggressiveness in their daily lives (Liebert *et al.*, 1977, pp. 330-33). The research has some inadequacies, but it is substantial enough to warrant concern (Gardner, 1978, pp. 322-26).

The kinds of models presented in the mass media have been of concern to those seeking social change, such as blacks and women's groups. These activists have pointed out that films, books, and magazines have typically depicted women and minorities in stereotyped roles and, by doing so, have restricted people's sense of what they might become in life. Accordingly, these groups have tried to get the media to present new kinds of models, such as women and blacks as doctors and scientists, rather than housewives and delinquents. The social learning research suggests that the activists have, in this case, adopted a good strategy for social change.

Because modeling can have a strong impact on behavior, it has significant promise as a therapeutic device. You might recall that in Mary Cover Jones's (1924) famous experiment, modeling was part of the method used to eliminate Peter's fear of furry objects. Bandura and others have conducted a number of studies that have more systematically shown how modeling can help reduce fears. In one experiment (Bandura *et al.*, 1967), for example, four-year-olds who were afraid of dogs observed a child calmly play with one, and then the children themselves became less fearful. Modeling can also help in other ways, such as making overly submissive clients more assertive (Rosenthal, 1976). In general, modeling has become one of the most popular innovations in the behavior modification area.

EVALUATION

The concept of imitation is an ancient one, but Bandura has greatly increased our awareness of its importance. He has shown how models can rapidly influence a wide range of behavior and has thus given us a better appreciation of how the social environment alters behavior. Furthermore, Bandura and his colleagues have made their points through a number of well-designed experiments, employing elegant and rigorous methodologies which are often lacking in developmental research.

At the same time, because social learning theorists have concentrated so heavily on laboratory research, we cannot yet be certain of how well their principles and findings hold up in ordinary life situations. For example, although Bandura showed that models can change children's moral thinking in the laboratory, we cannot be sure that modeling influences account for the growth of moral thinking in real life.

In some cases, the experimental situations may contain artificial conformity pressures that make the modeling effects look greater than they really are. In a typical experiment, the child is taken from the classroom and led into a room where he or she meets unfamiliar adults. Naturally, the child feels a bit overwhelmed and searches for clues as to the "right" way to behave. So, for example, if the child sees a model praised for discussing morals in terms of consequences, the child infers that he or she is supposed to do the same (see Cowan *et al.*, 1969).

Social learning theorists are aware of this problem and often do what they can to eliminate any artificial conformity pressures. Furthermore, it may well be that the conformity demands in experimental situations are no greater than those in ordinary life. Ultimately, then, the test will be how well the experimental findings hold up in more naturalistic research (Bandura and Walters, 1963, p. 42).

For us, the most important issues are those that divide Bandura and the developmentalists. To developmentalists, Bandura is too environmentalistic. When, for example, it comes to educational advice, Bandura asks us to consider how we teach by example. To developmentalists, this advice, while important, still diverts our attention away from the child. Bandura still has us looking to ourselves, or to other external agents, as the molders of the child's mind. Instead, developmentalists believe, we should study children, who, at different stages, may have strong interests that are very different from ours. If we are open to children's unique and spontaneous interests, we might then find out what they are most eager and ready to learn (see Montessori, 1949, Ch. 15).

Most developmentalists, it should be stressed, do not deny that the external environment does exert a considerable influence on us all, and often in ways that Bandura describes. However, they also want to make some room for growth that comes from the child, and they feel that Bandura too readily dismisses such growth.

Among the developmentalists, it has been the Piagetians who have become most embroiled in debates with Bandura. Kohlberg (1969a), for example, contends that Bandura underestimates the extent to which children spontaneously learn from an intrinsic interest in moderately novel events. Bandura (1977, p. 164) argues that the principle of moderate novelty does not fit with everyday observations; children, he says, really learn in order to obtain reinforcements, such as praise, which they eventually administer to themselves. However, Bandura's own research may contradict his argument. In several experiments, models perform "moderately novel" or "relatively unique" behaviors (e.g., Bandura, 1962, p. 220, 252; 1965b, p. 116), such as socking Bobo dolls, marching about, knocking objects off shelves, and other zany antics. These behaviors seem intuitively designed to capture the imagination of the four-year-old subjects, and in several experiments the subjects readily imitated them even though there were no reinforcements available (Bandura and Huston, 1961;

Bandura *et al.,* 1963; Bandura, 1965b). Quite possibly, they reproduced such behaviors because they found them intrinsically interesting. Reinforcement variables, to be sure, can increase or alter imitation, but at times spontaneous imitation may also be at work.

Secondly, many Piagetians (e.g., Kuhn, 1974) believe that Bandura ignores the importance of cognitive structures or stages. Bandura, as we have seen, does acknowledge that cognitive skills set limits on what children can learn, but he has given very little attention to such matters.

Perhaps some of Bandura's followers will be more inclined to consider the child's cognitive level in their research. In one interesting study, Liebert *et al.* (1969) found that 14-year-olds, but not eight- or six-year-olds, could imitate a new grammatical rule. The oldest subjects were able to figure out the rule underlying the model's behavior because they had the capacities for abstract thinking which the younger children lacked. They had, it seems, formal operations. There would appear to be countless opportunities for similar research, in which one could study the effects of models as mediated by cognitive capacities.

It may be, then, that Bandura's environmental theory underestimates the importance of developmental variables. It overlooks, that is, the extent to which children learn on their own and the extent to which modeling is influenced by the child's current stage. Nevertheless, Bandura has contributed substantially to our knowledge of how environmental forces operate.

14

Chomsky's
Theory of Language
Development[1]

BIOGRAPHICAL INTRODUCTION

In 1949 Montessori tried to get us to see that the child's mastery of language is an amazing achievement. It was not the learning of words that impressed her, but the acquisition of grammar or syntax—a system of rules for creating and understanding correct sentences. Grammatical rules are so complex and so deeply buried within spoken language that adults are hardly aware of them, yet children somehow unconsciously master most of them by the age of six years. Developmental psychology must understand how this happens.

Psychologists might have agreed with Montessori, but for a long time they were handicapped by their own ignorance of grammatical rules and structures. They were largely limited to counting children's nouns and verbs. Then, in 1957, Chomsky published *Syntactic Structures,* in which he described some of the operations we use to form and transform sentences. Researchers then had an idea of the kinds of operations to look for in children's speech, and the whole new field of developmental psycholinguistics emerged.

[1] This chapter was written in collaboration with Stephen Crain.

Noam Chomsky was born in 1928 in Philadelphia. He learned something about linguistics from his father, who was a respected Hebrew scholar, and studied linguistics, mathematics, and philosophy at the University of Pennsylvania. Actually, he was bored by college after two years and was ready to drop out when he met the linguist Zellig Harris. He became absorbed in Harris's work and then began making his own innovations. In the meantime, Chomsky earned both a B.A. and a Ph.D. degree from Pennsylvania, although he spent several graduate school years in a special program at Harvard which permitted him to work on whatever he wanted. Chomsky's new theory, a combination of mathematics and linguistics, was so different from anything that had been done before that the universities had no place for him within their traditional departments. His only job offer came from the Massachusetts Institute of Technology (MIT), where he has been since 1955 (Cohen, 1977, p. 80).

Chomsky is known not only for his innovations in linguistics but for his political activities and writings. He grew up in what he has called a "radical Jewish community" (Lyons, 1970, p. xi) and became one of the first intellectuals to speak out against the Vietnam War. He has also written extensively on American foreign policy. Many of his colleagues disagree with his radical politics, but they almost unanimously recognize his accomplishments as a linguist. He has been awarded two honorary doctorate degrees (from the University of Chicago and the University of London) and has received many other forms of recognition.

BASIC CONCEPTS

The Importance of Rules

Prior to Chomsky, most people probably believed in what Brown has called the "storage bin" theory of language learning. Children imitate others and acquire a large number of sentences which they store in their heads. They then reach in for the appropriate sentence when the occasion arises (Brown and Herrnstein, 1975, p. 444).

Chomsky has shown that this view is incorrect. We do not simply learn a set number of sentences, for we routinely create new ones. As I write this book, I use the same words over and over, but I create a novel sentence practically every time. We all do the same thing when we speak or write. We can do this because we have internal *rules* which enable us to decide which sentences are grammatical and convey our intended meanings. If we could only use sentences that we had already heard and memorized, our language would be severely limited. Because we have a system of rules—a grammar—we can invent and understand sentences that we have never heard before (Chomsky, 1959, p. 56).

Transformational Rules

Chomsky focuses on the rules for making transformations, as when we transform an active, declarative sentence into a question. Transformational rules are by no means obvious. Consider transforming these sentences into simple yes/no questions:

(1) The man is tall—Is the man tall?
    The book is on the table—Is the book on the table?
    I can go—Can I go?

The rule would appear to be to read from left to right until we come across the first verbal element ("is," "can," etc.), and then to move it to the front of the sentence. We can call this Hypothesis 1, and it works with almost every sentence of this type (Chomsky, 1975, p. 30). However, look what happens when we apply Hypothesis 1 to the following, similar sentence:

(2) The man who is tall is in the room—Is the man who tall is in the room?

Hypothesis 1 produces an ungrammatical question; Hypothesis 1, then, is wrong.

When we correctly transform sentences, we first analyze them into abstract phrases (e.g., the noun phrase and the verb phrase). The phrases are called "abstract" because there is nothing marking off their boundaries; our sense of them is intuitive. Then, in sentences of the above type, we locate the first verbal element ("is," "can," etc.) *after the first main noun phrase*, and it is this that we move to the front of the sentence. Thus,

(1)	The man	is	tall
	The book	is	on the table
	I	can	go
(2)	The man who is tall	is	in the room[2]

In this way, then, Sentence 2 is correctly transformed. Instead of moving to the front the first "is" we encounter, as Hypothesis 1 says, we move the first "is" after the first *entire* noun phrase ("The man who is tall"). We thereby produce the well-formed sentence, "Is the man who is tall in the room?" In Chomsky's terms, we follow a *structure-dependent* rule; we first analyze the structure of the phrases before making the transformation (Chomsky, 1975, p. 32).

How do we learn to correctly transform sentences like Sentence 2? It does not appear that we are taught how to do so, or even that our knowledge comes from any specific experience. We may live several years before we encounter a sentence like Sentence 2 which forces us to choose between the simpler Hypothesis 1 and the structure-dependent rule. All our experience may

[2] This diagram is suggested by Aitchison, 1976, p. 24.

be consistent with the simpler Hypothesis 1. And yet, upon encountering Sentence 2, we unhesitatingly apply the structure-dependent rule (Chomsky, 1975, p. 32).

What seems to happen is this. Early on, before children can make transformations, they become sensitive to phrase structures. Roger Brown's research suggests that as soon as children begin putting three or more words together, their pauses indicate an awareness that words fit into larger noun-phrase units. For example, the child says, "Put . . . the red hat . . . on," rather than "Put the . . . red hat on" (Brown and Bellugi, 1964, p. 150). Then, when children begin making transformations, they respect the integrity of these units.

Perhaps the child's sense of noun-phrase units is somehow learned, through conditioning or imitation. However, children appear to create these structures quite instantaneously, before conditioning could take place, and they do not appear to be imitating adult patterns of pausing (Aitchison, 1976, p. 208). More research is needed on these points, but it seems that the child's sense of noun-phrase units, the first building-block of a structure-dependent grammar, develops spontaneously.

Moreover, in *all* languages, the transformational rules are always structure-dependent. Theoretically, languages could employ various other systems, but they do not; they always follow structure-dependence. A very real possibility, Chomsky says, is that this is so because languages partly reflect the biological nature of the human mind. The mind is innately disposed to structure linguistic experience into units called phrases and to employ only structure-dependent transformational rules (Chomsky, 1975, pp. 32-35).

Surface and Deep Structure

When we create, comprehend, and transform sentences, we intuitively work on two levels. We attend both to the surface structure and to the deep structure of sentences. Chomsky's use of these concepts gets very abstract and technical (utilizing symbolic notation), but we can get a flavor of his ideas through some examples. Consider the following two sentences:

(3) John is eager to please.
(4) John is easy to please.

On the surface, the two sentences have the same structure: subject (John), predicate (is), modifier (eager/easy), and infinitive (to please). Yet we intuitively know that their underlying meanings, revealed by their deeper structures, are different. In Sentence 3, John is the subject; he is the one who pleases. In Sentence 4, John is the object; others please him. Thus, the two sentences have different underlying structures.

The surface/deep structure distinction is important with respect to transformations. Consider these sentences:

(5) Susan ate the apple.

(6) The apple was eaten by Susan.

(7) Susan did not eat the apple.

(8) What did Susan eat?

(9) Susan ate the apple, didn't she?

Of these sentences, Sentence 5 is the most straightforward. It is a simple, active, declarative sentence and follows the subject-verb-object word order. In English, this form is closest to (but not identical with) the abstract deep structure; Sentence 5 is the basic kernel upon which one can perform certain operations to generate all the other sentences. One could not take any other sentence—say Sentence 6—and derive a clear set of operations for creating the others (Chomsky, 1957, p. 45, 91; 1965, pp. 138-41).

Although the rules for making these transformations are too complex for us to go into here, children begin making strong headway toward mastering them by the age of five or six years. Children seem to ferret out the deep structure of their language first, and they then begin mastering transformations. This is not to say that they ever become consciously aware of transformational rules; even Chomsky is still trying to make them explicit. But they gain a working knowledge of them on an implicit level (McNeill, 1970).

A Language-Acquisition Device

Chomsky repeatedly emphasizes the amazing speed with which young children master complex linguistic rules. They learn the rules of their own language, and, if need be, those of a second language as well. It is a common observation that a young child of immigrant parents may learn a second language in the streets, from other children, with amazing rapidity, and the child may speak the new language as fluently as the other children (Chomsky, 1959, p. 42).

Chomsky acknowledges that everyone's speech, including that of adults, contains errors, slips, false starts, slurred expressions, fragmentary sentences, and so on. But these deficits in *performance* are far outweighed by an underlying *competence*, which is best revealed by an ability to distinguish between poorly formed and well-formed sentences (Chomsky, 1962, p. 531).

Chomsky (1959) says that the linguistic accomplishments of the ordinary child are too great to be explained solely in terms of reinforcement, imitation, and natural inquisitiveness. Children hear only a limited number of sentences, many of which are poorly formed, and yet they somehow develop sophisticated rules for creating and understanding an unlimited number of new sentences. To do this, they must be endowed with a special, innate language-acquisition device, or LAD. That is, they must be endowed with a mechanism which prompts them to raise hypotheses, search out regularities, and form rules—all of a special kind (e.g., only structure-dependent rules). The LAD is the human equivalent of the innate schematizing tendencies of the bird which enable it to learn various songs, as long as they fit into a certain structure. The LAD, which at this point is merely a rough hypothesis, is probably species-specific (found only in humans) and part of the mental apparatus of every physically normal child. It must be sufficiently structured to account for universal linguistic operations, yet not so highly structured as to be inconsistent with the known diversity in the world's languages. It is not fully formed at birth but grows with the maturation of the central nervous system (Chomsky, 1965, pp. 27-28, 54-58; 1975, Ch. 1).

NOTES ON THE GROWTH OF GRAMMAR

Although Chomsky has made general points with respect to language development, he has not carried out any empirical research on this topic himself. However, his work has inspired a growing number of psychologists to look at the ways children develop grammatical rules. Most have studied the speech of a few children over time. They have done this by tape recording children's spontaneous speech as unobtrusively as possible. The following notes summarize only some of the main findings.[3]

Early Language

Right at birth, babies seem tuned into language. Careful film analyses suggest that infants make very slight body movements in response to speech, and their movements vary with the boundaries of sounds and words. Such movements are not made in response to other sounds, such as tapping (Condon and Sander, 1974).

At two or three months of age, infants begin babbling, making sounds such as "ba ba ba" and "da da da." Early babbling appears to be identical throughout the world, probably because of the nature of the vocal apparatus in the early phases of development (Sachs, 1976, p. 148).

One-Word Utterances

At about one year, babies begin producing single words. Many researchers believe that they are trying to use single words to express entire sentences. For instance, "milk" might mean, "I want more milk," or "There is the milk," depending on the context (Bloom, 1970, p. 10).

[3] This summary basically follows that outlined by Cairns and Cairns, 1976, pp. 193-97.

Two-Word Utterances

Beginning at about one and a half years, children put two words together, and their language takes on a definite structure. Table 14.1 lists some typical two-word utterances. Notice how, in utterances 6 through 8, children are separating out agents, actions, and objects. They seem to be forming subject-verb-object relationships.

TABLE 14.1   Some Typical Two-Word Utterances

Type	Example
1. Naming	that doggie
2. Repetition	more jump
3. Negation	allgone ball, no wet
4. Possession	my truck
5. Attribution	big boy
6. Agent-action	Johnny hit
7. Action-object	hit ball
8. Agent-object	Mommy bread (meaning, "Mommy is cutting the bread")
9. Question	where Daddy

Adapted from Slobin, D. I. *Psycholinguistics.* Copyright © 1971 by Scott, Foresman and Co., pp. 44-45. And from Brown, R., and Herrnstein, R. J. *Psychology.* Copyright © 1975 by Little, Brown and Co., Inc., p. 478. By permission.

Developing Grammar

Between two and three years or so, children present a clearer sense of sentence structure. When the child uses a third word, it usually fills in the part that was implicit at the two-word stage. Sometimes we can actually see this happen, when the child changes an utterance in midstream. For example: "Want that . . . Andrew want that." At this point, the subject-verb-object sequence is all there (Slobin, 1971, p. 47; 1972). English-speaking children usually insist on following this precise word order, perhaps because it is integral to deep structure in our language (Brown *et al.*, 1969, p. 42).

It is during this period that children also begin putting words into larger units, especially noun phrases, as we noted earlier. They say, "Put . . . the red hat . . . on," or "The red hat . . . lost," indicating by their pauses that several words function as a unit (Brown and Bellugi, 1964, p. 150; Aitchison, 1976, p. 118).

During this time, in addition, children also begin using *inflections,* or word endings. When they do so, they *overregularize,* saying things such as "I runned," "It goed," and "She doed it." A similar process occurs with respect to plurals (e.g., "foots," "mans," and "mouses"). Overregularizations may persist well into the elementary school years (Slobin, 1972; Ervin, 1964).

What children seem to be doing is searching for general rules. They discover that the rule for forming the past tense is to add the "-ed" sound and then do so in all cases, assuming that the language is more consistent than it actually is. Similarly, they induce that the rule for creating plurals is to add the "-s" sound.

Children do not, however, just overgeneralize everything they hear. English-speaking children do not overgeneralize the "ing" ending. They say, "I making coffee," but not "I wanting." They see that the rule here is that verbs that describe actions, like "make," take the "-ing" ending, but verbs that describe states, like "want," do not. What seems utmost in children's minds, then, is the rule, and they overregularize when they think they have discovered one (Brown *et al.,* 1969, pp. 44-46).

At both the two-word phase and the present one, children use *negations*— for example, "no ball," "no drop mitten," and "get car no" (Dale, 1972, p. 69). For a time, children put the "no" either at the beginning or at the end of the sentence but not in the middle. This usage may reflect an unconscious understanding, once again, of deep structure, in which negatives stand outside the rest of the sentence and are used for making transformations (Klima, 1964).

Transformations

Between about three and six years, children's grammar rapidly becomes quite complex; most notably, children begin making transformations. Bellugi-Klima (1968) has studied how children form Where, What, and Why questions, which are transformations of their deep structure representations. For example, "Where can I put it?" is essentially a transformation of "I can put it where." Children do not master the transformational operations all at once. Typically, they first say, "Where I can put it?" and "What he wants?" They correctly move the "Where" or "What" to the front of the sentence, but they fail to invert the subject and the auxiliary verb. Even when asked to imitate, they commonly fail to make this inversion. For example,

Adult: "Adam, say what I say: Where can I put them?"

Adam: "Where I can put them?" (Slobin, 1971, p. 54)

Perhaps English-speaking children are reluctant to tamper with the subject-verb-object word order, the order which is basic to the deep structure which they have just mastered.

A particularly well-examined transformation is the creation of *tag questions.*

For example, when Brown's subject Adam was four and a half, he produced a number of tag questions in an hour of spontaneous speech. These included:

Ursula's my sister, isn't she?

I made a mistake, didn't I?

Diandros and me are working, aren't we?

He can't beat me, can he?

He doesn't know how to do it, does he?

The tags are the little questions on the ends of the sentences (Brown and Herrnstein, 1975, p. 471).

The creation of tag questions requires several operations. First of all, the speaker must reverse the negative or affirmative statement in the first part of the sentence. When Adam says, "I made a mistake," an affirmative statement, the tag must be negative, "Didn't I?" When he says, "He can't beat me," a negative statement, the tag must be positive, "Can he?" Adam also must locate the subject of the sentence (e.g., "Diandros and me") and convert it into the correct pronoun in the tag ("we"). Furthermore, he must use the proper auxiliary verb in the tag question. In the sentence "I made a mistake, didn't I," the auxiliary "did" was not even present in the initial phrase. Adam created it according to a complicated rule that we were not even aware of until Chomsky uncovered it (1957). And these are only some of the operations involved. It is truly remarkable that Adam can produce correct tag questions at age four and a half, yet he is not unusual in this respect.

An interesting aspect of Adam's verbal record is the frequency distribution of these sentences over time. Adam produced no tag questions at all until he was four and a half, and then he suddenly burst out with them. In one hour, he created 32 such questions, whereas the average adult frequency is three to six. His behavior is reminiscent of Piaget's circular reactions and Montessori's repetitions. Children seem to have an inner need to solidify new capacities through repeated exercise.

Near Adult Grammar

Although children master most aspects of grammar by the age of five or six years, some of the most complex transformations are still beyond their grasp. For example, they seem to have difficulty with the passive voice until age seven or so (Turner and Rommetveit, 1967). The years five to 10 may be important for the acquisition of the subtlest and most complex grammatical skills (C. Chomsky, 1969).

Universals

Children in different cultures learn to speak different languages. Nevertheless, Chomsky and his followers believe that amidst all the variety there may be certain universals to the developmental process. Children everywhere first seem to proceed from one-word to two-word utterances; they next work on rules with respect to inflections, deep structure, phrase structure units, and other elements; and, finally, they begin making transformations (McNeill, 1970; Slobin, 1972). Thus, children everywhere seem to learn in a standard sequence, perhaps a sequence which reflects the complexity of the tasks involved (Brown and Herrnstein, 1975, p. 481).

In addition, investigators are finding cross-cultural similarities at each general phase. Babbling, as mentioned, is the same throughout the world. Further, children everywhere may use the same kinds of two-word utterances, overgeneralize some parts of speech, and initially handle negatives in the same manner (Slobin, 1972, 1973; Cairns and Cairns, 1976, p. 205).

By the time children are working on transformations, however, they are learning rules that vary somewhat from language to language. Nevertheless, Chomsky argues that *universal constraints* restrict the range of possible transformations. We have already mentioned one major constraint; all languages seem to employ only structure-dependent rules. Chomsky says that universal constraints facilitate the learning process. Children do not have to hypothesize every possible kind of transformational rule; they are innately disposed to search only for rules of a certain kind. If they had to entertain and decide among a variety of possible transformational systems, it is unlikely that they would ever master language at all (Chomsky, 1975, p. 11).

CHOMSKY AND LEARNING THEORY

Chomsky suggests that language is something constructed by children themselves. Hearing only a fragmentary body of speech, they nevertheless dig out its structure and rules in a regular sequence, as their capacities mature. Learning theorists, in contrast, believe that we must look to the social environment for the source of linguistic patterning. Language, in their view, is primarily shaped by others, through their modeling influences and reinforcement practices.

Bandura and Modeling

Bandura emphasizes the influence of models. He recognizes that modeling does not always work through a process of exact imitation, for children produce novel utterances that they have never heard. For example, children's overregularizations (e.g., "We runned") cannot be exact imitations, for adults do not talk this way.[4] However, Bandura contends that modeling is still at work; the process is one of "abstract modeling." Children imitate the rules they hear (e.g., add the "-ed" sound to form the past tense), which they imitate too well (Bandura, 1977, p. 175).

The abstract modeling proposal has a certain plausibility in the case of overregularizations. However, children also create some linguistic structures which are unlike anything that adult models ever exemplify. For example, at one point children place the "no" exclusively at the beginning or end of sentences (and, later, use many double and triple negatives)—structures which adults may never model. Perhaps Bandura would contend that this is still modeling, but he would have to admit that the adult modeling influences are incredibly indirect.

In a very general sense, it is of course true that children learn language by hearing it from others. But when Bandura speaks of modeling influences, he implies more than this; he implies that what models are doing has a direct bearing on what children reproduce. Some laboratory research, in fact, suggests that the careful presentation of modeled utterances can influence children's speech (Whitehurst and Vasta, 1975). However, more naturalistic studies, which examine what children hear and say in their daily lives, have produced surprisingly little evidence that modeling influences are very powerful (Dale, 1972, pp. 91, 116-20; Bloom *et al.,* 1974). If, for example, adult models were all-important, one would expect to find children first using the forms of speech that adults use with the greatest frequency. However, they do not. Instead, children's grammar seems to follow its own developmental schedule of increasing complexity and elaboration (Brown and Herrnstein, 1975, p. 480).

Chomsky and the psycholinguists generally believe that imitation and modeling must exert some effect. However, for them, as for the developmentalists, there is one major problem with an environmental theory of modeling. It tells us to spend our time observing what models are doing. Instead, we would be better advised to learn more about children's own developing linguistic structures.

Operant Conditioning

The Skinnerian theory of language learning, which we already have discussed to some extent, is sometimes called the "babble luck" theory. Babies babble away until, by luck, they hit on a sound which resembles a word, and it is reinforced. For example, they say "da da" in the presence of Daddy, and the

---

[4] It is conceivable that overregularizations are exact imitations of the speech of other children, but it is unlikely. Overregularizations occur in all children, including two-year-old first-borns whose primary linguistic models are their parents (Slobin, 1971, pp. 49-50).

parents show their approval. Gradually, parents make their approval contingent upon increasingly accurate and complex utterances.

Skinner (1957) and his followers have recognized that such meticulous shaping of each utterance would be too slow a process to account for the rapid development of language. Accordingly, they have shown that when children are taught specific linguistic behaviors, they rapidly generalize their learning to new situations. For example, a child who has been taught to pluralize one word automatically pluralizes new words without any further training. Most of this research has been done with retarded children or with disturbed children who are behind in their speech. These children have been taught plurals, prepositions, and other relatively simple grammatical elements (Lovaas, pp. 110-16). However, it has not been shown that direct teaching can produce anything like complex grammatical transformations.

Furthermore, it is unlikely that normal children acquire language through parental conditioning, because parents are such poor language teachers. Brown and Hanlon (1970) found that parents correct surprisingly few ungrammatical utterances. Instead, they concentrate on the truthfulness of their children's remarks. For example, when one girl, Sarah, said, "Her curl my hair," her mother said, "That's right," because that was what she was doing. The mother ignored Sarah's grammatical error. When, on the other hand, Sarah said, "There's the animal farmhouse," a grammatically impeccable sentence, her mother corrected her, because it was a lighthouse (1970, p. 202). Thus, it is unlikely that Sarah learned grammar as a consequence of her parents' effective use of approval and disapproval.

Perhaps it is not parental approval, but some other form of feedback which constitutes effective reinforcement. Perhaps children learn to use increasingly correct grammar because parents can comprehend and respond accurately to it. However, Brown and Hanlon also failed to find that well-formed utterances met with any better understanding than poorly formed utterances. Brown and Hanlon's data were based on only three children, but these three seemed to learn correct grammar despite the poorest kind of direct reinforcement from their parents.

Recently, some of the most energetic learning theory research has been devoted to teaching sign language to apes. By teaching language to apes, learning theorists hope to demonstrate the effectiveness of their procedures and to refute Chomsky's assertion that language learning is an innate propensity specific to humans. Some researchers have relied on operant techniques alone, whereas others have mixed their methods. The results, in many cases, have been quite remarkable, but it is not yet clear that one can teach apes much beyond what children ordinarily can do at the two-word phase (Brown and Herrnstein, 1975, pp. 486-92).

The debates between the learning theorists and the psycholinguists will go on. At present, learning theorists have not shown that their variables govern the growth of children's language, but, as we shall see, neither have the psycholinguists produced much evidence with respect to maturational mechanisms. In the meantime, the controversies seem beneficial, for they are inspiring enthusiastic research.

IMPLICATIONS FOR EDUCATION

Chomsky says that children learn an extensive, intricate grammatical system almost entirely on their own. All they need is to hear a language spoken, and they will master it, without any special training program. This is as true of the child in the ghetto as it is of the child in the middle-class suburb. The child picks up language off the streets, without anyone paying much attention to his or her progress. In fact, it is often the immigrant child in a poor neighborhood who masters not just one language but two (Chomsky in Cohen, 1977, p. 88).

Chomsky and others repeatedly emphasize the magnitude of the ordinary child's accomplishment. As the Russian scholar Chukovsky has put it,

> It is frightening to think what an enormous number of grammatical forms are poured over the poor head of the young child. And he, as if it were nothing at all, adjusts to all this chaos, constantly sorting into rubrics the disorderly elements of the words he hears, without noticing as he does this, his gigantic effort. If an adult had to master so many grammatical rules within so short a time, his head would surely burst. . . . In truth, the young child is the hardest mental toiler on our planet (Chukovsky, 1963, p. 10).

Since Chomsky believes that children learn language spontaneously, he proposes no new instructional methods. This does not mean, however, that his work has no practical value. It can help change our attitudes, and deepen our appreciation of children's independent achievements. The teacher who considers the ghetto child's linguistic accomplishments will realize how ridiculous it is to focus on the child's shortcomings. Whatever the child may lack, it is trivial in comparison to the complex grammatical system he or she has mastered. The teacher, upon meeting each new elementary school child, will think: "Here is someone who has made monumental achievements. This mind deserves my greatest respect." One can only wonder about the effect such an attitude would have.

Despite the work of Chomsky and others, many psychologists cannot accept the possibility that children really learn language on their own. Instead, they believe that it is up to us to teach children a proper grammar. For example, Bandura (1977, p. 175) implies that we should correct children's overregularizations (e.g., "We digged the hole"). Chomsky's work suggests otherwise. What children are doing is searching for underlying rules, a search that will eventually lead to the mastery of an intricate grammar. It is wrong to interfere with this process. By correcting children's mistakes, we only confuse them and undermine their confidence. Their mistakes will correct themselves in time. Instead of correcting children, it would be more appropriate simply to admire their efforts. Our quiet admiration and joy can be as helpful as any corrections we might make.

Nevertheless, many educators are devising elaborate programs for teaching language—even to preschoolers. Frequently, their goal is to teach children with dialects such as Black English the standard English form. These educators commonly assume that Black dialects are inferior to standard English, which they are not (Bartlett, 1972; Labov, 1970). The likely outcome is that the black child is made to feel deficient and inadequate, feelings which can affect the child's attitudes in school.

Some of Chomsky's followers have also been interested in ways parents can effectively teach language to children. In particular, they have investigated parents' expanded imitations of children's speech. For example, one of Brown's subjects, Adam, said, "There go one," and his mother responded, "Yes, there goes one" (Brown and Bellugi, 1964, p. 140). Children then sometimes imitate their parents' expansions. Parents seem naturally to talk to children in this way, and children may enjoy such conversations. However, it is not at all clear that expansions are necessary or even beneficial for language acquisition (Brown, 1973, p. 329; Dale, 1972, pp. 116-20).

In schools, teachers describe the parts of speech, writing the familiar tree-diagrams on the blackboard. Before Chomsky's work, we might have thought that these lessons were necessary. However, the work of Chomsky and others suggests that children already have an implicit grasp of almost everything the teacher is explaining by the age of four years. Since tree-diagrams are abstract, children may benefit from them in adolescence, when the capacity for abstract thinking emerges. Before this time, they probably just baffle the child.

In general, the lesson to be gained from Chomsky's work is this: Since children independently master an intricate system of grammatical rules, we should respect their independent efforts. It is presumptuous of us to try to structure the child's learning, and our attempts to do so are likely to lead only to the child's loss of faith in his or her own intuitions. While it is good to talk to children in ways they find enjoyable, it is not necessary to do anything that undermines their confidence or depreciates their immense accomplishments.

EVALUATION

It is remarkable that Chomsky, who is more of a linguist or even a philosopher than a psychologist, has inspired so much psychological research. This research is a testament to the importance of his ideas.

We have focused on the descriptive studies of children's emerging grammar.

At the same time, Chomsky has stimulated other lines of investigation. Some researchers have been pursuing Chomsky's hypothesis that there is an innate neurological mechanism underlying language acquisition. For example, they have been examining evidence for a critical period for language which is related to cerebral lateralization. Once the left hemisphere, which usually controls language, has gained its specific dominance, it may be impossible for children to learn language with anything like the ease with which they do so before this time. Also, if children suffer brain damage, they may be able to recover linguistic functions much more readily if they can do so before the critical period for language is over. Some investigators think the outer limit for a critical period is age five, while others suggest the onset of puberty, but the evidence is still inconclusive (Krashen, 1975).

Chomsky has also stimulated research by posing a challenge to the Piagetians. Chomsky and Piaget, to be sure, have much in common. Both believe that children spontaneously grasp structures and operations—through their own efforts, not as a result of adult teachings. However, Chomsky proposes that language acquisition is an autonomous, maturationally governed process. It develops fairly independently from other kinds of cognitive development (Cohen, 1977, p. 90). Piagetians, in contrast, have suggested that language structure actually rests on prior cognitive achievements. For example, Sinclair (1971) thinks that the basic grammatical relations—subject-predicate-object—depend on sensori-motor developments. At first, babies fail to distinguish between themselves, their actions, and objects. If a girl drops a ball she thinks it is gone because it is no longer a part of her own actions. At the end of the sensori-motor period, children's search for missing objects implies a differentiation between themselves, their various actions, and the object itself. Thus, they develop the cognitive requirements for separating the self (subject), action (predicate), and object.

Because Piaget says that children do not think systematically again until the age of seven or so, when they attain concrete operations, Piagetians argue that children cannot perform many linguistic transformations until this time. For example, they cannot fully understand passive transformations, because these require reversible operations. Similarly, they cannot understand a sentence such as "The wolf is easy to bite," because they would need to reverse the sentence into something like, "Someone can easily bite the wolf" (Cromer, 1970; Ingram, 1975). In this Piagetian view, then, concrete operations are necessary before basic linguistic transformations are possible.

Chomsky's followers would counter that, while it is true that some difficult transformations may develop after age six, children also develop many complex transformations before six—during a period when Piaget suggests that they have no logical operations. Consequently, language itself cannot be entirely determined by general cognitive development. The task of research, then, is to understand just how much children can do linguistically before concrete operations emerge.

A great deal of research will be necessary before we can make any balanced judgments with respect to Chomsky's strong claims. If there is any criticism of Chomsky at present, it is his tendency to stay above the research battles. He often writes as if he could settle the various debates through logical argument alone. He is, as we mentioned, more a philosopher and logician than a psychologist.

In concluding, it is worth recalling how much Montessori anticipated Chomsky. She suggested that children acquire linguistic rules unconsciously under the influence of an innate language acquisition mechanism which unfolds during a critical period and contributes to universal linguistic structures. Her own position was based primarily on hunches—on observations that children seem to be learning so many complex rules in so short a time that they must be innately disposed to do so. Chomsky's work has given us a more precise idea of what kinds of rules they are mastering, and therefore it has stimulated much interesting research into how this happens.

15

Conclusion:
Humanistic Psychology
and Developmental Theory

In this, the concluding chapter, we will discuss the emergence of humanistic psychology and indicate the extent to which developmental theorists have shared the humanists' concerns.

HUMANISTIC PSYCHOLOGY

Background

Humanistic psychology is only a single recent expression of an old intellectual tradition, a tradition with roots in ancient Greek, Hebrew, and Asian philosophy. There really is no one set of humanistic beliefs; humanism has taken different forms in different historical eras. It has emerged whenever people have felt that some system or authority—political, moral, or intellectual—was undermining human dignity or human unity (Fromm, 1967; Hawton, 1961).

Much early humanistic writing was directed against religious dogma. For example, sixteenth-century writers such as Desiderius Erasmus and Sir Thomas More protested that the Church often asked people to blindly adhere to religious

256

doctrines, thus undermining their dignity by robbing them of the freedom to think for themselves. Furthermore, the Church often pitted Christians against non-Christians, when it should have been promoting tolerance and brotherhood (de Santillana, 1956, pp. 27-30, 92-95).

Humanistic philosophy flourished most widely during the eighteenth-century Enlightenment, when the writings of Locke struck such a responsive cord. Before Locke, political and religious authorities often argued that people were innately wicked and therefore required repression. But if Locke were right—if people were solely the products of their environments—then one had only to change the environments to perfect them and make repression unnecessary. Furthermore, if inequalities were not innate but the products of circumstances, one could erase these too.

The Enlightenment spirit, then, was quite optimistic; unlimited human progress seemed possible. To achieve the greatest progress, Locke, Voltaire, and others turned to science. They believed that the scientific method—the open-minded, cautious verification of hypotheses through empirical evidence—would produce the knowledge that would better the condition of all (Berlin, 1956, pp. 11-29).

The Enlightenment beliefs in science and environmentalism have continued to play a role in some humanistic thinking. However, no single set of beliefs has ever claimed the complete loyalty of all humanists. In fact, in modern psychology, the combination of science and environmentalism took on a form which began to strike some as too rigid and narrow, leading to a new humanistic revolt.

Modern Scientific Psychology

Modern scientific psychology grew out of a deep admiration for the achievements of physics, chemistry, and the other natural sciences. Many psychologists felt that if psychology could only follow the example of these sciences, it too could accomplish great things. They proposed that psychology, like physics, strive for the objective, quantitative measurement of isolated variables and the formulation of abstract laws. At first, in the latter part of the nineteenth century, scientific psychology was led by Wundt and his colleagues, who tried to analyze consciousness into basic elements, as physicists and chemists had done. This effort, however, seemed to lead to a dead end, and by the second decade of the twentieth century, the scientific banner had been taken over by another group, the behaviorists. The behaviorists, as we have seen, argued that we should confine ourselves to the measurement of overt behavior and the way it is controlled by observable stimuli in the external environment. The inner world of thoughts, feelings, and fantasies, they argued, has little place in scientific psychology.

Many behaviorists have seen themselves within the humanistic tradition; they have been trying to develop scientific techniques that will better the human lot. As we have seen, experimenters have used behavior modification techniques to alleviate fears, temper tantrums, and other problems. They also have devised new instructional methods. Behaviorists, like their Enlightenment predecessors, have held that their environmental, scientific approach will produce the greatest human progress.

The Humanistic Revolt in Psychology

Early on, however, some psychologists had misgivings about the behavioristic brand of science. During the first half of this century, Gordon Allport, Carl Rogers, Abraham Maslow, and others argued that behaviorism, whatever its merits, was producing a very one-sided picture of human nature. Humans, they argued, do not consist only of overt responses, nor are they completely controlled by the external environment. People also grow, think, feel, dream, create, and do many other things which make up the human experience. The behaviorists and others, in their emulation of the physical sciences, were ignoring most aspects of life which make humans unique and give them dignity. These humanists were not at all opposed to scientific investigation, but they argued that psychology should address itself to the full range of human experience, not just the aspects that are most readily measurable and under environmental control. For some time, these writers were calling out in the wilderness; their views were far removed from the mainstream in American psychology. But in the 1950s, their writings began to attract increasing attention, and a humanistic movement in psychology was born (Misiak and Sexton, 1973, pp. 108-9).

The humanistic psychologists' call for renewed attention to inner experience has aligned them with the existential and phenomenological movements. With the existentialists, the humanists have held that the living person must take priority over any abstract, scientific system. With the phenomenologists, the humanists have pointed to the need to suspend our ordinary ways of classifying people from the outside; instead we should try to understand how the world feels to people from the inside.

Modern humanistic psychology, then, primarily developed in reaction to behavioristically oriented approaches. Humanistic psychology's relationship to the second main branch of psychology, psychoanalysis, has been more ambivalent. Many humanists have appreciated the psychoanalytic attempt to explore the inner world at its deepest levels. However, humanists have also felt that the psychoanalysts have been too pessimistic about human capacities for growth and free choice. Whereas the behaviorists have seen people as exclusively controlled by the external environment, psychoanalysts have viewed people as dominated by irrational forces in the unconscious. Perhaps, humanists have suggested, psychoanalytic theory has been too colored by the study of patients with crippling emotional disorders. Humanists have proposed that people, to a much greater extent than has been realized, are free and creative beings, capable of growth and self-actualization (Maslow, 1962, pp. 189-97).

The modern humanistic movement in psychology, then, sees itself diverging from the two dominant forces in psychology, scientific behaviorism and psychoanalysis. For this reason, humanists sometimes call their movement the "third force." To get a more concrete picture of the kind of work humanistic psychologists have been doing, we will now review some of the ideas of the man who is considered the father of modern humanistic psychology, Abraham Maslow.

Maslow

Biographical introduction. Maslow (1908-1970) was born in Brooklyn, New York, the son of poor, Russian immigrant parents. He was a shy, unhappy boy. Although he liked high school, he had trouble adjusting to college. He attended the City College of New York, Cornell University, and finally the University of Wisconsin, where he earned his B.A. degree and stayed on for graduate work in psychology. Maslow began his career squarely within the scientific mainstream. He received rigorous experimental training under E. L. Thorndike and Harry Harlow and wrote a standard textbook on abnormal psychology (Wilson, 1972, pp. 115-34). In fact, Maslow said that early in his career he was sold on behaviorism (Misiak and Sexton, 1973, p. 113), and in a sense he never repudiated it; he always realized that people are subject to conditioning from the external environment. What increasingly annoyed him was behaviorism's one-sidedness; people also have an inner life and potentials for growth, creativity, and free choice.

Maslow's ideas. Maslow's first step in the direction of a humanistic psychology was the formulation of a new theory of motivation (1943). According to this theory, there are six kinds of needs: physiological needs, safety needs, belongingness needs, love needs, self-esteem needs, and, at the highest level, self-actualization needs. These needs are arranged in a hierarchical order such that the fulfillment of lower needs propels the organism on to the next highest level. For example, a man who has a strong physiological need, such as hunger, will be motivated by little else, but when this need is fulfilled, he will move on to the next level, that of safety needs, and when these are satisfied, he will move on to the third level, and so on.

In his major works, Maslow was most interested in the highest need, the need for self-actualization. Self-actualization, a concept borrowed from Goldstein (1939), refers to the actualization of one's potentials, capacities, and talents. To study this concept, Maslow examined the lives and experiences of the most healthy, creative people he could find. His sample included both contemporaries and acquaintances, such as the anthropologist Ruth Benedict, as well as public and historical figures, such as Thomas Jefferson and Eleanor Roosevelt (1954, pp. 202-3).

Maslow's key finding was that the self-actualizers, compared to most people, have maintained a certain independence from their society. Most people are so strongly motivated by needs such as belongingness, love, and respect that they are afraid to entertain any thought that others might disapprove of. They try to fit into their society and do whatever brings prestige within it. Self-actualizers, in contrast, are less conforming. They seem less molded and flattened by the social environment and are more spontaneous, free, and natural. Although they rarely behave in unconventional ways, they typically regard conventions with a good-natured shrug of the shoulders. Instead, they are primarily motivated by their own inner growth, the development of their potentials, and their personal mission in life (1954, pp. 223-28).

Because self-actualizers have attained a certain independence from their culture, they are not confined to conventional, abstract, or stereotyped modes of perception. When, for example, most people go to a museum, they read the name of the artist below the painting and then judge the work according to the conventional estimate. Self-actualizers, in contrast, perceive things more freshly, naively, and as they really are. They can look at any painting—or any tree, bird, or baby—as if seeing it for the first time; they can find miraculous beauty where others see nothing but the common object (Maslow, 1966, p. 88). In fact, they seem to have retained the creative, open approach which is characteristic of the young child. Like the child, their attitude is frequently "absorbed, spellbound, popeyed, enchanted" (1966, p. 100).[1] Unfortunately, most children lose this approach to life as they become socialized.

In a sense, then, self-actualizers are good phenomenologists. They can suspend or go beyond conventional ways of ordering experience. Maslow also likened their approach to a "Taoistic letting be," to an appreciation of objects without interfering with them or attempting to control them (1962, p. 86). He suggested that psychologists can learn much from self-actualizers' phenomenological and Taoistic approaches. Instead of fitting people into theoretical categories and performing experiments on them, psychologists should first simply watch and listen to people with a naive openness. If they did this, they would be open to new discoveries.

Maslow reworked his ideas over the years and was not always systematic in the process. But by and large, his overall position was as follows:

---

[1] When such perception is intense, it can be called a "peak experience." Peak experiences may emerge during almost any activity—not only when observing nature, but during work, artistic production, love, or athletics. During a peak experience, people lose themselves in the moment and things seem to flow and to come almost without striving or effort. People are at the peak of their powers (Maslow, 1962, Chs. 6 and 7).

1. Humans possess an essential, biological, inner nature, which includes all the basic needs and the impulses toward growth and self-actualization (1962, p. 190; 1971, p. 25).

2. This inner core is partly species-wide and partly idiosyncratic, for we all have special bents, temperaments, and abilities (1962, p. 191).

3. Our inner core is a positive force which presses toward the realization of full humanness, just as an acorn may be said to press toward becoming an oak tree. It is important to recognize that it is our inner nature, not the environment, that plays the guiding role. The environment is like sun, food, and water; it nourishes growth, but it is not the seed. Social and educational practices should be evaluated not in terms of how efficiently they control the child or get the child to adjust, but according to how well they support and nourish inner growth potentials (1962, pp. 160-61, 211-12).

4. Our inner nature is not strong, like instincts in animals. Rather, it is subtle, delicate, and in many ways weak. It is easily "drowned out by learning, by cultural expectations, by fear, by disapproval, etc." (1962, p. 191).

5. The suppression of our inner nature usually takes place during childhood. At the start, babies have an inner wisdom with respect to most matters, including food intake, amount of sleep, readiness for toilet-training, and the urges to stand up and to walk. Babies will also avidly explore the environment, focusing on the particular things in which they take delight. Their own feelings and inner promptings guide them toward healthy growth. However, socializing agents frequently lack respect for children's choices. Instead, they try to direct children, to teach them things. They criticize them, correct their errors, and try to get them to give the "right" answers. Consequently, children quit trusting themselves and their senses and begin to rely on the opinions of others (1962, pp. 49-55, 150, 198-99).

6. Even though our inner core, with its urge toward self-actualization, is weak, it rarely disappears altogether—even in adulthood. It persists underground, in the unconscious, and speaks to us as an inner voice waiting to be heard. Inner signals can lead even the neurotic adult back to buried capacities and unfulfilled potentials. Our inner core is a pressure we call the "will to health," and it is this urge on which all successful psychotherapy is based (1962, pp. 192-93; 1971, p. 33).

DEVELOPMENTALISTS AS HUMANISTS

If Maslow's ideas sound familiar, they are. Maslow and the modern humanistic psychologists have, without making much note of it, drawn heavily upon the developmental tradition. Since Rousseau, developmentalists have been preoccupied with the same basic problem as Maslow: Children, as they become socialized, quit relying on their own experience and judgments; they become too dependent on conventions and the opinions of others. Thus, the developmentalists, like the humanists, have been searching for an inner force that will guide the individual toward a healthier, more independent development.

Intrinsic Growth Forces

Where Maslow speaks of a biological core which directs healthy growth, developmentalists refer to maturation. Maturation is an internal mechanism which prompts children to seek out certain experiences at certain times. Under maturational urging, children regulate their cycles of sleep and eating, learn to sit up, walk, and run, develop an urgent need for autonomy, master language, explore the widening environment, and so on. According to Gesell and others, children, following their own inner schedule and timing, are eminently wise regarding what they need and can do. So, instead of trying to make children conform to our own set schedules and directions, we can let them guide us and make their own choices—as Maslow proposed.

Nevertheless, as Maslow observed, it is often difficult for us to trust children and the growth process. We seem to have particular difficulty believing that children can really learn on their own, without our direction and supervision. But developmentalists have tried to show that they can. Montessori, in particular, tried to show that if we will open-mindedly observe children's spontaneous interests, they will direct us to the tasks on which they will work independently and with the greatest concentration and sense of fulfillment. They will become absorbed in such tasks because the tasks meet inner needs to perfect certain capacities at certain points in development. Thus, we are not forced to take charge of children's learning, to choose tasks for them, to motivate them by our praise, or to criticize their mistakes—practices which force them to turn to external authorities for guidance and evaluation. Instead, we can trust their maturationally based urges to perfect their own capacities in their own ways. Maslow might have pointed to Montessori as an educator who was thoroughly humanistic in her faith in children's intrinsic creative powers.

Not all developmentalists, of course, are as nativistic as Gesell or Montessori. As we have seen, Piaget, Kohlberg, and the cognitive-developmentalists doubt that biological maturation directly governs the stages of cognitive development. Nevertheless, these theorists also look to children's independent activities, rather than to external teachings, as the source of development. Children, in their view, are intrinsically curious about the world and reach out for new experiences that lead them to reorganize their cognitive structures. Thus, the cognitive-developmentalists also share the humanists' faith in intrinsic capacities for self-directed learning.

Interestingly, Maslow's thoughts on adulthood also were foreshadowed by earlier developmental theorists—especially by Jung. Maslow pointed out how the well-socialized adult, whose inner potentials for self-actualization lie dormant, will still hear inner voices calling for attention. Jung used nearly identical language to describe the crisis of middle life. Prior to middle age, the individual typically concentrates on adjusting to the external, social world. The man or woman tries to do things that bring social success and prestige and develops those parts of the personality that are suited for this goal. In middle life, however, social success loses its importance, and inner voices from the unconscious direct one to attend to the previously neglected and unrealized parts of the self. The individual increasingly turns inward and considers the discovery and rounding out of the personality more important than social conformity.

Thus, developmental theorists, like the modern humanistic psychologists, have tried to uncover intrinsic growth factors that stand apart from pressures toward social conformity. At the same time, however, some developmental theorists have been more pessimistic than the humanists about the chances for any substantial improvement based on intrinsic forces. In particular, the Freudians have felt that because maturation brings with it unruly sexual and aggressive impulses, a good measure of social repression will always be necessary. Erikson has viewed maturational growth somewhat more positively than Freud, calling attention to the maturation of autonomy, initiative, industry, and so on, but he too has felt that the other sides of these strengths—shame, doubt, guilt, inferiority, and so on—are inevitable. No child, for example, can become completely autonomous, for societies will always need to regulate the child to some extent. Still, Erikson hopes that we can raise children so that they can gain as much autonomy, initiative, and as many other virtues as possible.

Furthermore, Freudian therapy relies heavily on inner growth forces. Recall how Freud once asked a psychiatrist if he could really cure. When the psychiatrist responded that he could not—that he could only remove some impediments to growth like a gardener removes some stones or weeds—Freud said that they would then understand each other. The psychoanalyst's reliance on intrinsic growth processes is quite evident in Bettelheim's school. Bettelheim does not try to make disturbed children behave in normal ways, but he tries to provide certain conditions—love, acceptance, empathy—which will enable children to feel it is safe to take steps toward growth on their own. The physician treats patients in essentially the same way. The doctor does not actually heal a cut but only cleans and stitches the wound. The rest is up to Nature. Any cure, in psychotherapy or in medicine, partly relies on forces toward health that are out of the doctor's control. The doctor puts his or her faith in innate forces toward health.

Thus, developmental theorists, like the humanists, have tried to discover the nature of intrinsic growth forces and to devise educational and therapeutic methods based upon them. And, to a considerable extent, developmental writers had been working on these tasks long before the modern humanistic movement in psychology even began.

Phenomenology

Another central component of modern humanistic psychology is a phenomenological orientation. This orientation or method includes what may be called a "phenomenological suspension." One tries to suspend one's theoretical preconceptions and customary categories and tries to see people and things as openly and freshly as possible—to see them as they really are. This approach, as we have seen, was the starting point of Rousseau's developmental philosophy. Rousseau argued that children have their own ways of seeing, thinking, and feeling, and that we know nothing about these; we therefore must refrain from investing children with our own thoughts and take the time to simply observe them, listen to them, and let them reveal their unique characteristics to us. Later, Piaget and Montessori emphasized the same point. The ethologists, too, may be said to employ a phenomenological suspension. Before an ethologist forms any hypothesis or builds any theory, he or she first simply tries to learn about and describe as much about a particular species as possible. To do this, ethologists believe, we must observe animals in their natural habitats, not in the laboratory.

In psychology, phenomenology usually implies a second step. Phenomenological psychologists usually suspend preconceptions in order to enter into the *inner* world of the other. They try to open themselves to the other's direct experience, to see things through the other's eyes.

Developmental theorists have been less consistent in taking this second step. Those who have worked the hardest to learn about children's inner worlds are Schachtel and the psychoanalysts. Schachtel tried to gain insight into the infant's unique modes of perception, and the psychoanalyst Bettelheim, for example, constantly asks himself, "How does the world look and feel to this child?" Other writers, however, have been less interested in perceiving the world through the child's eyes. Gesell wanted us to be open to children's own needs and interests, but he primarily observed their external motor behavior. Piaget asks us to understand the child's thought in its own terms, but he too examines it from the outside, analyzing it in terms of logical structures. The ethologists also look primarily at external behavior.

A knowledge of how the world looks to children (and adults) at different stages will not be easy to come by. Young children are not sufficiently verbal to tell us how the world appears to them, and infants cannot tell us anything at all. One approach may be the study of spontaneous interests. For example, Montessori showed how young children attend to minute details and are concerned about anything out of place. These observations give us two clues concerning the young child's perceptual world. Young children also seem to perceive life where we do not, and they may be particularly interested in objects, such as cars, balls, or balloons, which, with a little imagination, take on human qualities. It would seem important to record every aspect of the environment that seems uniquely interesting to children. Such a record could contribute to a developmental phenomenology.

Universals

Readers who have already learned something about developmental psychology will notice that this book neglects or skims over certain topics. We have not discussed, for example, differences among children or adults on I.Q. tests. Other topics that have received only minor coverage include cultural differences in personality development and sex differences. The various differences among people, which are partly the product of environmental factors, are tremendously important. If we are ever to redress artificial social inequalities, we need to know how they are formed.

However, the differences among people have not been the developmentalists' primary concern. Instead, they have searched for developmental forces and sequences common to all peoples. This search, as Chomsky suggests (1975, pp. 130-33) probably reflects, as much as anything, an ethical orientation. Developmentalists, like humanists, are trying to show how, at the deepest levels, we are all the same. Developmentalists shy away from topics concerning how one person is better than another and focus, instead, on what Maslow calls our "biological brotherhood" (1962, p. 185). They want to show that, at bottom, we all have the same yearnings, hopes, and fears, as well as the same creative urges toward health and personal integration. Hopefully, an appreciation of the positive strivings that we all share can help in the building of a universal human community.

SUMMARY

Humanistic thought has taken various forms through the decades. During the Enlightenment, humanism became linked to an environmental outlook and the scientific enterprise. The hope was that the creation of improved environments, aided by scientific knowledge, would lead to better living conditions for us all. In contemporary psychology, the behaviorists continue to pursue the goals of the Enlightenment.

In recent years, however, the behavioristic approach has struck many humanistic psychologists as too one-sided. Maslow and others have objected that behaviorism leaves out too much that gives human life its richness and dignity. By focusing on how external behavior comes under environmental control, it has ignored our inner world and our spontaneous urges toward health and independence. If we are to create better environments, they must not simply be those that control behavior but those that foster and support the intrinsic creative forces.

In formulating a new version of humanism, Maslow and others have, without making much note of it, drawn heavily upon the developmental tradition in psychology. Most notably, they have joined the developmentalists in their search for intrinsic growth forces that can lead to healthy development. Also, the developmentalists have, to an extent, anticipated the humanists' call for a more phenomenological science; for the developmentalists have suggested that we cannot appreciate the growing child's unique ways of thinking and learning unless we approach the child with something of a naive openness. Finally, both humanists and developmentalists have been most concerned with positive universals, with the impulses toward growth and health that are part of our universal human nature.

Bibliography

ABRAHAM, K. (1924a). A short study of the development of the libido viewed in light of mental disorders. *Selected Papers of Karl Abraham.* New York: Basic Books, 1927.

—— (1924b). The influence of oral eroticism on character formation. *Selected Papers of Karl Abraham.* New York: Basic Books, 1927.

AINSWORTH, M. D. S. (1962). The effects of maternal deprivation: A review of findings and controversy in the context of research and strategy. *Public Health Papers, 14.* Geneva: World Health Organization.

—— (1967). *Infancy in Uganda: Infant Care and the Growth of Love.* Baltimore: Johns Hopkins University Press.

—— (1973). The development of infant and mother attachment. In B. M. Caldwell and H. M. Ricciuti (Eds.), *Review of Child Development Research* (Vol. III). Chicago: University of Chicago Press.

AITCHISON, J. (1976). *The Articulate Mammal: An Introduction to Psycholinguistics.* New York: University Books.

ALMY, M., CHITTENDEN, E., and MILLER, P. (1966). *Young Children's Thinking.* New York: Columbia Teachers' College Press.

AMES, L. B. (1971). Don't push your preschooler. *Family Circle Magazine, 79,* 60.

APPLETON, T., CLIFTON, R., and GOLDBERG, S. (1975). The development of behavioral competence in infancy. In F. D. Horowitz (Ed.), *Review of Child Development Research* (Vol. IV). Chicago: University of Chicago Press.

ARIÈS, P. (1960). *Centuries of Childhood: A Social History of Family Life* (R. Baldick, trans.). New York: Knopf, 1962.

ARNHEIM, R. (1954). *Art and Visual Perception.* Berkeley: University of California Press.

AUSUBEL, D. P. (1958). *Theories and Problems in Child Development.* New York: Grune & Stratton.

BALDWIN, A. L. (1967). *Theories of Child Development.* New York: John Wiley.

BALINSKY, B. I. (1970). *An Introduction to Embryology* (3rd ed.). Philadelphia: Saunders.

BANDURA, A. (1962). Social learning through imitation. In M. R. Jones (Ed.), *Nebraska Symposium on Motivation.* Lincoln, Neb.: University of Nebraska Press.

—— (1964). The stormy decade: Fact or fiction? *Psychology in the School, 1,* 224-31.

—— (1965a). Vicarious processes. A case of no-trial learning. In L. Berkowitz (Ed.), *Advances in Experimental Social Psychology* (Vol. II). New York: Academic Press.

—— (1965b). Influence of model's reinforcement contingencies on the acquisition of imitative responses. *Journal of Personality and Social Psychology, 1,* 589-95.

—— (1967). The role of modeling processes in personality development. In W. W. Hartup and W. L. Smothergill (Eds.), *The Young Child: Reviews of Research.* Washington, D.C.: National Association for the Education of Young Children.

—— (1969). Social-learning theory of identificatory processes. In D. A. Goslin (Ed.), *Handbook of Socialization Theory and Research.* Chicago: Rand McNally.

—— (1971). Analysis of modeling processes. In A. Bandura (Ed.), *Psychological Modeling.* Chicago: Atherton, Aldine.

—— (1977). *Social Learning Theory.* Englewood Cliffs, N.J.: Prentice-Hall.

——, GRUSEC, J. E., and MENLOVE, F. L. (1967). Vicarious extinction of avoidance behavior. *Journal of Personality and Social Psychology, 5,* 16-23.

——, and HUSTON, A. C. (1961). Identification as a process of incidental learning. *Journal of Abnormal and Social Psychology, 63,* 311-18.

——, and KUPERS, C. J. (1964). The transmission of patterns of self-reinforcement through modeling. *Journal of Abnormal and Social Psychology, 69,* 1-9.

——, and McDONALD, F. J. (1963). Influence of social reinforcement and the behavior of models in shaping children's moral judgments. *Journal of Abnormal and Social Psychology, 67,* 274-81.

——, ROSS, D., and ROSS, S. A. (1963). A comparative test of the status envy, social power, and secondary reinforcement theories of identificatory learning. *Journal of Abnormal and Social Psychology, 67,* 527-34.

——, and WALTERS, R. H. (1963). *Social Learning and Personality Development.* New York: Holt, Rinehart, & Winston.

BAERENDS, G., BEER, C., and MANNING, A. (1975). *Function and Evolution in Behavior.* Oxford: Clarendon Press.

BARTLETT, E. J. (1972). Selecting preschool language programs. In C. B. Cazden (Ed.), *Language in Early Childhood Education.* Washington, D.C.: National Association for the Education of Young Children.

BATESON, P. P. G. (1966). The characteristics and context of imprinting. *Biological Reviews, 41,* 177-220.

BAUMRIND, D. (1967). Child care practices anteceding three patterns of preschool behavior. *Genetic Psychology Monographs, 75,* 43-88.

BELL, S. M. (1970). The development of the concept of object as related to infant-mother attachment. *Child Development, 41,* 291-311.

——, and AINSWORTH, M. D. S. (1972). Infant crying and maternal responsiveness. *Child Development, 43,* 1171-90.

BELLUGI-KLIMA, U. (1968). Linguistic mechanisms underlying child speech. In E. M. Zale (Ed.), *Proceedings of the Conference on Language and Language Behavior.* Englewood Cliffs, N.J.: Prentice-Hall.

BENEDEK, T. (1938). Adaptation to reality in early infancy. *Psychoanalytic Quarterly, 7,* 200-15.

BERES, D. (1971). Ego autonomy and ego pathology. *Psychoanalytic Study of the Child, 26,* 3-24.

BERGMAN, I. (1957). *Wild Strawberries* (filmscript). (L. Malmstrom and D. Kushner, trans.). New York: Simon & Schuster.

BERLIN, I. (1956). *The Age of Enlightenment: The Eighteenth Century Philosophers.* New York: Mentor.

BETTELHEIM, B. (1960). *The Informed Heart: Autonomy in a Mass Age.* New York: Free Press.

—— (1967). *The Empty Fortress: Infantile Autism and the Birth of the Self.* New York: Free Press.

—— (1974). *A Home for the Heart.* New York: Knopf.

—— (1976). *The Uses of Enchantment: The Meaning and Importance of Fairy Tales.* New York: Knopf.

BIJOU, S. W. (1976). *Child Development: The Basic Stage of Early Childhood.* Englewood Cliffs, N.J.: Prentice-Hall.

——, and BAER, D. M. (1961). *Child Development* (Vol. I). Englewood Cliffs, N.J.: Prentice-Hall.

BIRCH, H. G. (1956). Sources of order in the maternal behavior of animals. *American Journal of Orthopsychiatry, 26,* 279-84.

BLACKMAN, D. (1974). *Operant Conditioning: An Experimental Analysis of Behavior.* London: Methuen and Co., Ltd.

BLATT, M. M., and KOHLBERG, L. (1975). The effects of classroom moral discussions upon children's level of moral judgment. *Journal of Moral Education, 4,* 129-61.

BLOOM, L. (1970). *Language Development: Form and Function in Emerging Grammars.* Cambridge, Mass.: The MIT Press.

——, HOOD, L., and LIGHTBOWN, P. (1974). Imitation in language development: If, when, and why? *Cognitive Psychology, 6,* 380-420.

BLOS, P. (1962). *On Adolescence.* New York: Free Press.

BORKE, H. (1975). Piaget's mountains revisited: Changes in the egocentric landscape. *Developmental Psychology, 11,* 240-43.

BOWER, T. G. R. (1976). Repetitive processes in child development. *Scientific American, 235,* 38-47.

BOWLBY, J. (1953). *Child Care and the Growth of Love.* Baltimore: Penguin Books.

—— (1969). *Attachment and Loss* (Vol. I). *Attachment.* New York: Basic Books.

—— (1973). *Attachment and Loss* (Vol. II). *Separation.* New York: Basic Books.

BRACKBILL, Y. (1958). Extinction of the smiling response in infants as a function of reinforcement schedule. *Child Development, 29,* 115-24.

BREUER, J., and FREUD, S. (1895). *Studies in Hysteria* (A. A. Brill, trans.). New York: Nervous and Mental Disease Publishing Co., 1936.

BRONFENBRENNER, U. (1960). Freudian theories of identification and their derivatives. *Child Development, 31,* 15-40.

BROWN, J. F. (1940). *The Psychodynamics of Abnormal Behavior.* New York: McGraw-Hill.

BROWN, P., and ELLIOTT, R. (1965). Control of aggression in a nursery school class. *Journal of Experimental Child Psychology, 2,* 103-7.

BROWN, R. (1965). *Social Psychology.* New York: Free Press.

—— (1973). *A First Language: The Early Stages.* Cambridge, Mass.: Harvard University Press.

——, and BELLUGI, U. (1964). Three processes in the child's acquisition of syntax. *Harvard Educational Review, 34,* 133-51.

——, CAZDEN, C., and BELLUGI-KLIMA, U. (1969). The child's grammar from I to III. In J. P. Hill (Ed.), *Minnesota Symposia on Child Psychology* (Vol. II). Minneapolis: University of Minnesota Press.

——, and HANLON, C. (1970). Derivational complexity and order of acquisition in child speech. In Brown, R., *Psycholinguistics: Selected Papers.* New York: Free Press.

——, and HERRNSTEIN, R. J. (1975). *Psychology.* Boston: Little, Brown.

BRYAN, J. H. (1975). Children's cooperation and helping behaviors. In E. M. Hetherington (Ed.), *Review of Child Development Research* (Vol. 5). Chicago: University of Chicago Press.

——, and WALBEK, N. (1970). Preaching and practicing generosity: Children's action, and reactions. *Child Development, 41,* 329-53.

BUTLER, R. N. (1963). The life review: An interpretation of reminiscence in the aged. *Psychiatry, 26,* 65-76. Reprinted in B. L. Neugarten (Ed.), *Middle Age and Aging.* Chicago: University of Chicago Press, 1968.

CAIRNS, H. S., and CAIRNS, C. E. (1976). *Psycholinguistics: A Cognitive View of Language.* New York: Holt, Rinehart & Winston.

CALDWELL, B. M. (1964). The effects of infant care. In M. L. and L. W. Hoffman (Eds.), *Review of Child Development Research* (Vol. I). New York: Russell Sage Foundation.

CHANDLER, M. J., and GREENSPAN, S. (1972). Eratz egocentrism: A reply to H. Borke. *Developmental Psychology, 7,* 107-9.

CHOMSKY, C. (1969). *The Acquisition of Syntax in Children from 5 to 10.* Cambridge, Mass.: MIT Press.

CHOMSKY, N. (1957). *Syntactic Structures.* The Hague: Moulton.

—— (1959). A review of *Verbal Behavior* by B. F. Skinner. *Language, 35,* 26-58.

—— (1962). Explanatory models in linguistics. In E. Nagel, P. Suppes, and A. Tarshi (Eds.), *Logic, Methodology and Philosophy of Science.* Stanford: Stanford University Press.

—— (1965). *Aspects of the Theory of Syntax.* Cambridge, Mass.: The MIT Press.

—— (1975). *Reflections on Language.* New York: Pantheon.

CHUKOVSKY, K. (1963). *From Two to Five* (M. Morton, trans.). Berkeley: University of California Press, 1968.

COATES, B., and HARTUP, W. W. (1969). Age and verbalization in observational learning. *Developmental Psychology, 1,* 556-62.

COHEN, D. (1977). *Psychologists on Psychology.* New York: Taplinger.

COLES, R. (1970). *Erik H. Erikson: The Growth of His Work.* Boston: Little, Brown.

CONDON, W. S., and SANDER, L. W. (1974). Neonate movement is synchronized with adult speech: Interactional participation and language acquisition. *Science, 183,* 99-101.

CORMAN, H. H., and ESCALONA, S. K. (1969). Stages of sensori-motor development: A replication study. *Merrill-Palmer Quarterly, 15,* 351-61.

COWAN, P. A., LANGER, J., HEAVENRICH, J., and NATHANSON, J. (1969). Social learning and Piaget's theory of moral development. *Journal of Personality and Social Psychology, 11,* 261-74.

CROMER, R. F. (1970). "Children are nice to understand." Surface structure clues for the recovery of deep structure. *British Journal of Psychology, 61,* 397-408.

DALE, P. S. (1972). *Language Development: Structure and Function.* Hinsdale, Ill.: The Dryden Press.

DARWIN, C. (1859). *The Origin of Species.* New York: Modern Library.

—— (1871). *The Descent of Man.* New York: Modern Library.

—— (1887). *The Autobiography of Charles Darwin.* New York: W. W. Norton & Co., Inc., 1958.

DASEN, P. R. (1972). Cross-cultural Piagetian research: A summary. *Journal of Cross-Cultural Psychology, 3,* 23-39.

DAVIS, C. M. (1939). Results of the self-selection of diets by young children. *Canadian Medical Association Journal, 41,* 257-61.

DEWEY, J., and DEWEY, E. (1915). *Schools for Tomorrow.* New York: Dutton.

EHRLICH, P. R., and HOLM, K. W. (1963). *The Process of Evolution.* New York: McGraw-Hill.

EISENBERG, L., and KANNER, L. (1956). Early infantile autism, 1943-1955. *American Journal of Orthopsychiatry, 26,* 556-66.

ELLENBERGER, H. F. (1958). A clinical introduction to psychiatric phenomenology and existential analysis. In R. May, E. Angel, and H. F. Ellenberger (Eds.), *Existence: A New Dimension in Psychiatry and Psychology.* New York: Basic Books.

—— (1970). *The Discovery of the Unconscious.* New York: Basic Books.

ERIKSON, E. H. (1950). *Childhood and Society* (2nd ed.). New York: W. W. Norton & Co., Inc., 1963.

—— (1958). *Young Man Luther.* New York: W. W. Norton & Co., Inc.

—— (1959). Identity and the life cycle. *Psychological Issues* (Vol. I, No. 1). New York: International Universities Press.

—— (1964). *Insight and Responsibility.* New York: W. W. Norton & Co., Inc.

—— (1969). *Gandhi's Truth.* New York: W. W. Norton & Co., Inc.

—— (1976). Reflections on Dr. Borg's life cycle. *Daedalus,105,* 1-28.

ERVIN, S. M. (1964). Imitation and structural change in children's language. In E. H. Lenneberg (Ed.), *New Directions in the Study of Language.* Cambridge, Mass.: MIT Press.

ESTES, W. K. (1944). An experimental study of punishment. *Psychological Monographs, 57,* 94-107.

ETZEL, B. C., and GEWIRTZ, J. L. (1967). Experimental modification of care-taking maintained high-rate operant crying in a 6- and a 20-week-old infant (*infans tyrannotearus*): Extinction of crying with reinforcement of eye contact and smiling. *Journal of Experimental Child Psychology, 5,* 303-17.

EVANS, R. I. (1969). *Dialogue with Erik Erikson.* New York: Dutton.

FANTZ, R. L. (1961). The origin of form perception. *Scientific American, 204,* 459-63.

FENICHEL, O. (1945). *The Psychoanalytic Theory of Neurosis.* New York: W. W. Norton & Co., Inc.

FLAVELL, J. H. (1963). *The Developmental Psychology of Jean Piaget.* New York: Van Nostrand Reinhold.

—— (1977). *Cognitive Development.* Englewood Cliffs, N.J.: Prentice-Hall.

——, BOTKIN, P. T., FRY, C. L., WRIGHT, J. W., and JARVIS, P. E. (1968). *The Development of Role-Taking and Communication Skills in Children.* New York: John Wiley.

FREEDMAN, D. G. (1971). An evolutionary approach to research on the life cycle. *Human Development, 14,* 87-99.

—— (1974). *Human Infancy: An Evolutionary Perspective.* New York: John Wiley.

FREUD, A. (1936). *The Ego and the Mechanisms of Defense.* New York: International Universities Press, 1946.

—— (1958). Adolescence. *Psychoanalytic Study of the Child, 13,* 255-78.

FREUD, S. (1900). *The Interpretation of Dreams* (J. Strachey, trans.). New York: Basic Books (Avon), 1965.

—— (1905). Three contributions to the theory of sex. *The Basic Writings of Sigmund Freud* (A. A. Brill, trans.). New York: The Modern Library.

—— (1907). The sexual enlightenment of children (J. Riviere, trans.). *Collected Papers* (Vol. II). New York: Basic Books, 1959.

—— (1908a). Character and anal eroticism (J. Riviere, trans.). *Collected Papers* (Vol. II). New York: Basic Books, 1959.

—— (1908b). On the sexual theories of children (J. Riviere, trans.). *Collected Papers* (Vol. II). New York: Basic Books, 1959.

—— (1909). Analysis of a phobia in a five-year-old boy (A. and J. Strachey, trans.). *Collected Papers* (Vol. III). New York: Basic Books, 1959.

—— (1910). *The Origin and Development of Psychoanalysis.* New York: Henry Regnery (Gateway Editions), 1965.

—— (1911). Formulations regarding the two principles of mental functioning (J. Riviere, trans.). *Collected Papers* (Vol. IV). New York: Basic Books, 1959.

—— (1912). Contributions to the psychology of love: The most prevalent form of degradation in erotic life (J. Riviere, trans.). *Collected Papers* (Vol. IV). New York: Basic Books, 1959.

—— (1913). The excretory functions in psychoanalysis and folklore (J. Strachey, trans.). *Collected Papers* (Vol. V). New York: Basic Books, 1959.

—— (1914a). On the history of the psychoanalytic movement (J. Riviere, trans.). *Collected Papers* (Vol. I). New York: Basic Books, 1959.

—— (1914b). On narcissism: An introduction (J. Riviere, trans.). *Collected Papers* (Vol. IV). New York: Basic Books, 1959.

—— (1915a). Instincts and their vicissitudes (J. Riviere, trans.). *Collected Papers* (Vol. IV). New York: Basic Books, 1959.

—— (1915b). The unconscious (J. Riviere, trans.). *Collected Papers* (Vol. IV). New York: Basic Books, 1959.

—— (1916). Metapsychological supplement to the theory of dreams (J. Riviere, trans.). *Collected Papers* (Vol. IV). New York: Basic Books, 1959.

—— (1917). Mourning and melancholia (J. Riviere, trans.). *Collected Papers* (Vol. IV). New York: Basic Books, 1959.

—— (1920). *A General Introduction to Psychoanalysis* (J. Riviere, trans.). New York: Washington Square Press, 1965.

—— (1922). Medusa's head (J. Strachey, trans.). *Collected Papers* (Vol. V). New York: Basic Books, 1959.

—— (1923). *The Ego and the Id* (J. Riviere, trans.). New York: W. W. Norton & Co., Inc., 1960.

—— (1924). The passing of the Oedipus complex (J. Riviere, trans.). *Collected Papers* (Vol. II). New York: Basic Books, 1959.

—— (1925a). Some psychological consequences of the anatomical distinction between the sexes (J. Strachey, trans.). *Collected Papers* (Vol. V). New York: Basic Books, 1959.

—— (1925b). The resistance to psychoanalysis (J. Strachey, trans.). *Collected Papers* (Vol. V). New York: Basic Books, 1959.

—— (1931). Female sexuality (J. Strachey, trans.). *Collected Papers* (Vol. V). New York: Basic Books, 1959.

—— (1933). *New Introductory Lectures on Psychoanalysis* (J. Strachey, trans.). New York: W. W. Norton & Co., Inc., 1965.

—— (1936a). *The Problem of Anxiety* (H. A. Bunker, trans.). New York: The Psychoanalytic Press and W. W. Norton & Co., Inc.

—— (1936b). A disturbance in memory on the Acropolis (J. Strachey, trans.). *Collected Papers* (Vol. V). New York: Basic Books, 1959.

—— (1940). *An Outline of Psychoanalysis* (J. Strachey, trans.). New York: W. W. Norton & Co., Inc., 1949.

FROMM, E. (1967). Introduction. In E. Fromm (Ed.), *Socialist Humanism.* London: Allen Page, The Penguin Press.

GARDNER, H. (1973). *The Arts and Human Development.* New York: John Wiley.

—— (1978). *Developmental Psychology: An Introduction.* Boston: Little, Brown.

GELFAND, D. M. (1969). *Social Learning in Childhood.* Belmont, Calif.: Brooks/Cole.

GELMAN, R. (1969). Conservation acquisition: A problem of learning to attend to relevant attributes. *Journal of Experimental Child Psychology, 7,* 167-87.

GESELL, A. (1945). *The Embryology of Behavior.* New York: Harper and Row, Pub.

—— (1946). The ontogenesis of infant behavior. In L. Carmichael (Ed.), *Manual of Child Psychology* (2nd ed.). New York: John Wiley, 1954.

—— (1952a). Autobiography. In E. G. Boring, H. Werner, R. M. Yerkes, and H. Langfield (Eds.), *A History of Psychology in Autobiography* (Vol. IV). Worcester, Mass.: Clark University Press.

—— (1952b). *Infant Development: The Embryology of Early Human Behavior.* Westport, Conn.: Greenwood Press, 1972.

——, and AMATRUDA, C. S. (1941). *Developmental Diagnosis: Normal and Abnormal Child Development.* New York: Hoeber.

——, and ILG, F. L. (1943). *Infant and Child in the Culture of Today.* In A. Gesell and F. L. Ilg, *Child Development.* New York: Harper and Row, Pub., 1949.

——, and ILG, F. L. (1946). *The Child from Five to Ten.* In A. Gesell, and F. L. Ilg, *Child Development.* New York: Harper and Row, Pub., 1949.

——, and THOMPSON, H. (1929). Learning and growth in identical infant twins: An experimental study by the method of co-twin control. *Genetic Psychology Monographs, 6,* 1-124.

GINSBURG, H., and OPPER, S. (1969). *Piaget's Theory of Intellectual Development*. Englewood Cliffs, N.J.: Prentice-Hall.

GITELSON, M. (1975). The emotional problems of elderly people. In W. C. Sze (Ed.), *Human Life Cycle*. New York: Jason Aronson.

GLEASON, J. B. (1967). Do children imitate? *Proceedings of the International Conference on Oral Education of the Deaf*, June 17-24.

GOLDSTEIN, K. (1939). *The Organism: A Holistic Approach to Biology Derived From Pathological Data in Man*. New York: American Book.

GOUIN-DÉCARIE, T. (1965). *Intelligence and Affectivity in Early Childhood*. New York: International Universities Press.

GRIMM, THE BROTHERS (1972). *The Complete Grimm's Fairy Tales*. New York: Random House.

GRUSEC, J. E., and BRINKER, D. B. (1972). Reinforcement for imitation as a social learning determinant with implications for sex-role development. *Journal of Personality and Social Psychology, 21,* 149-58.

HAAN, N., SMITH, M. B., and BLOCK, J. (1968). Moral reasoning of young adults: Political-social behavior, family background, and personality correlates. *Journal of Personality and Social Psychology, 10,* 183-201.

HALL, C. (1954). *A Primer of Freudian Psychology*. New York: Mentor Book (New American Library).

——, and LINDZEY, G. (1975). *Theories of Personality* (2nd ed.). New York: John Wiley.

HARTMANN, H. (1939). *Ego Psychology and the Problem of Adaptation*. New York: International Universities Press, 1958.

—— (1950). Comments on the psychoanalytic theory of the ego. In H. Hartmann, *Essays on Ego Psychology*. New York: International Universities Press, 1964.

—— (1956). The development of the ego concept in Freud's work. *International Journal of Psychoanalysis, 37,* 425-38.

——, KRIS, E., and LOWENSTEIN, R. M. (1946). Comments on the formation of psychic structure. *Psychoanalytic Study of the Child, 2,* 11-38.

HAVIGHURST, R. J. (1952). *Developmental Tasks and Education*. New York: David McKay.

—— (1968). A social-psychological perspective on aging. *The Gerontologist, 8,* 67-71.

——, NEUGARTEN, B. L., and TOBIN, S. S. (1968). Disengagement and patterns of aging. In B. L. Neugarten (Ed.), *Middle Age and Aging*. Chicago: University of Chicago Press.

HAWTON, H. (1961). Humanism versus authoritarianism. In M. Knight (Ed.), *Humanist Anthology*. Bungay, Suffolk (Great Britain): Barrie & Rockliff.

HESS, E. H. (1962). Ethology: An approach toward the complete analysis of behavior. In *New Directions in Psychology* (Vol. I). New York: Holt, Rinehart & Winston.

—— (1973). *Imprinting: Early Experience and the Developmental Psychology of Attachment*. New York: Van Nostrand Reinhold.

HETHERINGTON, E. M., and PARKE, R. D. (1977). *Contemporary Readings in Child Psychology*. New York: McGraw-Hill.

HOFFMAN, M. L. (1970). Moral development. In P. H. Mussen (Ed.), *Carmichael's Manual of Child Psychology* (3rd ed., Vol. II). New York: John Wiley.

HOGAN, R. (1975). Theoretical egocentrism and the problem of compliance. *American Psychologist, 30,* 533-40.

HOLSTEIN, C. B. (1973). Irreversible, stepwise sequence in the development of moral judgment: A longitudinal evaluation. Paper presented at the biannual meeting of the Society for Research in Child Development, March, 1973.

HOLT, J. (1964). *How Children Fail.* New York: Dell Pub. Co., Inc.

HOMME, L. E., and TOTSI, D. T. (1965). Contingency management and motivation. *National Society for Programmed Instruction Journal, 4,* No. 7. Reprinted in D. M. Gelfand (Ed.), *Social Learning in Childhood: Readings in Theory and Application.* Belmont, Calif.: Brooks/Cole, 145-50.

HONIGMANN, J. J. (1967). *Personality in Culture.* New York: Harper & Row, Pub.

INGRAM, D. (1975). If and when transformations are acquired. In D. P. Dato (Ed.), *Developmental Psycholinguistics.* Washington, D.C.: Georgetown University Press.

INHELDER, B. (1971). The criteria of the stages of mental development. In J. M. Tanner and B. Inhelder (Eds.), *Discussions on Child Development.* New York: International Universities Press.

——, and PIAGET, J. (1955). *The Growth of Logical Thinking from Childhood to Adolescence* (A. Parsons and S. Milgram, trans.). New York: Basic Books, 1958.

JACOBI, J. (1965). *The Way of Individuation* (R. F. C. Hull, trans.). New York: Harcourt Brace Jovanovich, 1967.

JACOBSON, E. (1954). The self and the object world. *Psychoanalytic Study of the Child, 9,* 75-127.

JAHODA, G. (1958). Child animism. I. A critical survey of cross-cultural research. *Journal of Social Psychology, 47,* 197-212.

JONES, E. (1918). Anal-erotic character traits. *Journal of Abnormal Psychology, 13,* 261-84.

—— (1961). *The Life and Work of Sigmund Freud* (ed. and abridged by J. Trilling and S. Marcus). New York: Basic Books.

JONES, M. C. (1924). A laboratory study of fear: The case of Peter. *Pedagogical Seminary, 31,* 308-15.

JONES, N. B. (1972). *Ethological Studies of Child Behavior.* Cambridge, England: Cambridge University Press.

JUNG, C. G. (1931). Marriage as a psychological relationship (R. F. C. Hull, trans.). In C. G. Jung, *Collected Works* (Vol. XX). *The Development of Personality.* Princeton: Princeton University Press, 1953.

—— (1933). *Modern Man in Search of a Soul* (W. S. Dell and C. F. Baynes, trans.). New York: Harvest Book.

—— (1945). The relations between the ego and the unconscious. (R. F. C. Hull, trans.). *The Collected Works of C. G. Jung* (Vol. VII). *Two Essays in Analytic Psychology.* Princeton: Princeton University Press, 1953.

—— (1961). *Memories, Dreams, Reflections* (A. Jaffe, Ed., R. and C. Winston, trans.). New York: Vintage Books.

—— (1964). Approaching the unconscious. In C. G. Jung (Ed.), *Man and His Symbols.* New York: Dell Pub. Co., Inc.

KANNER, L. (1943) Autistic disturbances of affective contact. *Nervous Child, 2,* 217-50.

KANT, I. (1788). *The Critique of Practical Reason* (L. W. Beck, trans.). New York: Liberal Arts Press, 1956.

KARDINER, A. (1945). *The Psychological Frontiers of Society.* New York: Columbia University Press.

——, and PREBLE, E. (1961). *They Studied Man.* New York: Meridian Books.

KENISTON, K. (1971). The perils of principle. In Keniston, K., *Youth and Dissent.* New York: Harcourt Brace Jovanovich.

KESSEN, W. (1965). *The Child.* New York: John Wiley.

KIMBLE, G. A. (1961). *Hilgard and Marquis' Conditioning and Learning.* Englewood Cliffs, N.J.: Prentice-Hall.

KLIMA, E. S. (1964). Negation in English. In J. J. Fodor and J. A. Katz (Eds.), *The Structure of Language.* Englewood Cliffs, N.J.: Prentice-Hall.

KOHLBERG, L. (1958a). *The Development of Modes of Thinking and Choice in the Years 10 to 16.* Unpublished doctoral dissertation, The University of Chicago.

—— (1958b). *Global Rating Guide with New Materials.* School of Education, Harvard University.

—— (1963). The development of children's orientations toward a moral order: I. Sequence in the development of moral thought. *Vita Humana, 6,* 11-33.

—— (1964). Development of moral character and moral ideology. In M. L. Hoffman and L. W. Hoffman (Eds.), *Review of Child Development Research* (Vol. I). New York: Russell Sage Foundation.

—— (1966a). Cognitive stages and preschool education. *Human Development, 9,* 5-17.

—— (1966b). A cognitive-developmental analysis of children's sex-role concepts and attitudes. In E. E. Maccoby (Ed.), *The Development of Sex Differences.* Stanford: Stanford University Press.

—— (1968). Early education: A cognitive-developmental approach. *Child Development, 39,* 1013-62.

—— (1969a). Stage and sequence. A cognitive-developmental approach to socialization. In D. A. Goslin (Ed.), *Handbook of Socialization Theory and Research.* Chicago: Rand McNally.

—— (1969b). The relations between moral judgment and moral action. Colloquium presented at the Institute of Human Development. Berkeley: University of California Press.

—— (1970). The child as a moral philosopher. *Readings in Developmental Psychology Today.* Del Mar, Calif.: CRM Books.

——, and ELFENBEIN, D. (1975). The development of moral judgments concerning capital punishment. *American Journal of Orthopsychiatry, 45,* 614-40.

——, and GILLIGAN, C. (1971). The adolescent as philosopher. *Daedalus, 100,* 1051-86.

——, KAUFFMAN, K., SCHARF, P., and HICKEY, J. (1975). The just community approach to corrections: A theory. *Journal of Moral Education, 4,* 243-60.

——, and KRAMER, R. (1969). Continuities and discontinuities in childhood and adult moral development. *Human Development, 12,* 93-120.

KRAMER, R. (1976). *Maria Montessori: A Biography.* New York: Putnam's.

KRASHEN, S. D. (1975). The development of cerebral dominance and language learning: More new evidence. In D. P. Dato (Ed.), *Developmental Psycholinguistics: Theory and Applications.* Washington, D.C.: Georgetown University Press.

KUHN, D. (1974). Inducing development experimentally: Comments on a research paradigm. *Developmental Psychology, 10,* 590-600.

—— (1976). Short-term longitudinal evidence for the sequentiality of Kohlberg's early stages of moral judgment. *Developmental Psychology, 12,* 162-66.

KURTINES, W., and GRIEF, E. B. (1974). The development of moral thought: Review and evaluation of Kohlberg's approach. *Psychological Bulletin, 81,* 453-70.

LABOV, W. (1970). The logic of nonstandard English. In F. Williams (Ed.), *Language and Poverty: Perspectives on a Theme.* Chicago: Markham.

LAING, R. D. (1965). *The Divided Self: An Existential Study in Sanity and Madness.* Middlesex, England: Penguin.

—— (1967). *The Politics of Experience.* New York: Ballantine Books.

LEVINSON, D. (1977). The mid-life transition. *Psychiatry, 40,* 99-112.

—— (1978). *The Seasons of a Man's Life.* New York: Ballantine.

LIEBERT, R. M., ODOM, R. D., HILL, J. H., and HUFF, R. L. (1969). Effects of age and rule familiarity on the production of modeled language constructions. *Developmental Psychology, 1,* 108-12.

LIEBERT, R. M., POULOS, R. W., and MARMOR, G. S. (1977). *Developmental Psychology* (2nd ed.). Englewood Cliffs, N.J.: Prentice-Hall.

LILLARD, P. P. (1972). *Montessori: A Modern Approach.* New York: Shocken Books.

LIPSITT, L. P. (1975). The synchrony of respiration, heart rate, and sucking behavior in the newborn. *Biologic and Clinical Aspects of Brain Development,* Mead Johnson Symposium on Prenatal and Developmental Medicine, No. 6. Reprinted in Smart, R. C., and Smart, M. S. (Eds.), *Readings in Child Development and Relations* (2nd ed.). New York: Macmillan, 1977.

LOCKE, J. (1689). *Two Treatises on Government,* P. Laslett, (Ed.). Cambridge, England: Cambridge University Press, 1960.

—— (1690). *Essay Concerning Human Understanding* (Vol. I, J. W. Yolton, Ed.). London: J. M. Dent and Sons Ltd., 1961.

—— (1963). *Some Thoughts Concerning Education.* P. Gay (Ed.), *John Locke on Education.* New York: Bureau of Publications, Teacher's College, Columbia University, 1964.

LOMAX, E. M. R., KAGAN, J., and ROSENKRANTZ, B. G. (1978). *Science and Patterns of Child Care.* San Francisco. W. H. Freeman & Company Publishers.

LORENZ, K. (1935). Companions as factors in the bird's environment. In Lorenz, K. *Studies in Animal and Human Behavior* (Vol. I) (R. Martin, trans.). Cambridge, Mass.: Harvard University Press, 1971.

—— (1937). The establishment of the instinct concept. In Lorenz, K. *Studies in Animal and Human Behavior* (Vol. I) (R. Martin, trans.). Cambridge, Mass.: Harvard University Press, 1971.

—— (1952a). The past twelve years in the comparative study of behavior. In C. H. Schiller (Ed.), *Instinctive Behavior.* New York: International Universities Press, 1957.

—— (1952b). *King Solomon's Ring* (M. K. Wilson, trans.). New York: Thomas Y. Crowell.

—— (1963). *On Aggression.* New York: Harcourt Brace Jovanovich.

—— (1965). *Evolution and Modification of Behavior.* Chicago: University of Chicago Press.

LOVAAS, O. I. (1973). *Behavioral Treatment of Autistic Children.* University Programs Modular Studies. Morristown, N.J.: General Learning Press.

—— (1977). *The Autistic Child.* New York: Halstead Press.

——, SIMMONS, J. Q., KOEGEL, R. L., and STEVENS, J. (1973). Some generalization and follow-up measures on autistic children in behavior therapy. *Journal of Applied Behavior Analysis, 6,* 131-66.

LOVELL, K. (1968). Piaget in perspective: The experimental foundations. Paper presented to the conference of the University of Sussex. Sussex, England, April 5 and 6, 1968.

LOZOFF, B., BRITTENHAM, G. M., TRAUSE, M. A., KENNELL, J. H., and KLAUS, M. H. (1977). The mother-newborn relationship: Limits of adaptability. *The Journal of Pediatrics, 91,* 1-12.

LYONS, J. (1970). *Noam Chomsky.* New York: Viking Press.

MACCOBY, E. E., and WILSON, W. C. (1957). Identification and observational learning from films. *Journal of Abnormal and Social Psychology, 55,* 76-87.

——, and JACKLIN, C. N. (1974). *The Psychology of Sex Differences.* Stanford, Calif.: Stanford University Press.

McNEILL, D. (1970). The development of language. In P. H. Mussen (Ed.), *Carmichael's Manual of Child Psychology* (3rd ed., Vol. I). New York: John Wiley.

MALINOWSKI, B. (1927). *Sex and Repression in Savage Society.* New York: Harcourt Brace Jovanovich.

MARCIA, J. E. (1966). Development and validation of ego identity status. *Journal of Personality and Social Psychology, 3,* 551-58.

MARLER, P., and TAMURA, M. (1964). Culturally transmitted patterns of vocal behavior in sparrows. *Science, 146,* 1483-86.

MASLOW, A. (1943). A dynamic theory of human motivation. *Psychological Review, 50,* 370-96.

—— (1954). *Motivation and Personality* (2nd ed.). New York: Harper & Row, Pub., 1970.

—— (1962). *Toward a Psychology of Being* (2nd ed.). Princeton: D. Van Nostrand, 1968.

—— (1966). *The Psychology of Science: A Reconnaissance.* Chicago: Henry Regnery (Gateway), 1969.

—— (1971). *The Farther Reaches of Human Nature.* New York: Viking.

MEAD, M. (1928). *Coming of Age in Samoa.* New York: Mentor, 1949.

—— (1964). *Continuities in Cultural Evolution.* New Haven: Yale University Press.

MISCHEL, W. (1970). Sex-typing and socialization. In P. H. Mussen (Ed.), *Carmichael's Manual of Child Psychology* (3rd ed., Vol. II). New York: John Wiley.

MISIAK, H., and SEXTON, V. S. (1973). *Phenomenological, Existential, and Humanistic Psychologies: A Historical Survey.* New York: Grune & Stratton.

MONTESSORI, M. (1917). *The Advanced Montessori Method* (Vol. I), *Spontaneous Activity in Education* (F. Simmonds, trans.). Cambridge, Mass.: Robert Bentley, Inc., 1964.

—— (1936a). *The Child in the Family* (N. R. Cirillo, trans.). Chicago: Henry Regnery Co., 1970.

—— (1936b). *The Secret of Childhood* (M. J. Costelloe, trans.). New York: Ballantine Books, 1966.

—— (1948). *The Discovery of the Child* (M. J. Costelloe, trans.). Notre Dame, Ind.: Fides Publishers, 1967.

—— (1949). *The Absorbent Mind* (C. A. Claremont, trans.). New York: Holt, Rinehart & Winston, 1967.

—— (1970). *Maria Montessori: A Centenary Anthology, 1870-1970.* Koninginneweg, Amsterdam: Association Montessori Inernationale.

MUNN, N. L. (1974). *The Growth of Human Behavior* (3rd ed.). Boston: Houghton Mifflin.

——, FERNALD, L. D., and FERNALD, P. S. (1974). *Introduction to Psychology* (3rd ed.). Boston: Houghton Mifflin.

MUNROE, R. (1955). *Schools of Psychoanalytic Thought.* New York: Henry Holt and Co., Inc.

MUSSEN, P. H., CONGER, J. J. and KAGAN, J. (1974). *Child Development and Personality* (4th ed.). New York: Harper & Row, Pub.

——, and EISENBERG-BERG, N. (1977). *Roots of Caring, Sharing, and Helping.* San Francisco: W. H. Freeman & Company Publishers.

MUUSS, R. E. (1975). *Theories of Adolescence* (3rd ed.). New York: Random House.

NEEDHAM, J. (1959). *A History of Embryology* (2nd ed.). Cambridge, England: Cambridge University Press.

NEILL, A. S. (1960). *Summerhill: A Radical Approach to Child Rearing.* New York: Hart Publishing Co.

NEIMARK, E. D. (1975). Longitudinal development of formal operations thought. *Genetic Psychology Monographs, 91,* 171-225.

NEUGARTEN, B. L. (1964). A developmental view of adult personality. In J. E. Birren (Ed.), *Relations of Development and Aging.* Springfield, Ill.: Charles C Thomas.

—— (1968). Adult personality: Toward a psychology of the life cycle. In Neugarten, B. L. (Ed.), *Middle Age and Aging.* Chicago: University of Chicago Press.

PARSONS, A. (1964). Is the Oedipus complex universal? In W. Muensterberger and S. Axelrad (Eds.), *The Psychoanalytic Study of Society* (Vol. III). New York: International Universities Press.

PAVLOV, I. P. (1927). *Conditioned Reflexes* (G. V. Anrep, trans.). London: Oxford University Press.

—— (1928). *Lectures on Conditioned Reflexes* (Vol. I, W.H. Gantt, trans.). New York: International Publishers.

PEILL, E. J. (1975). *Invention and Discovery of Reality.* London: John Wiley.

PIAGET, J. (1923). *The Language and Thought of the Child* (M. Gabain, trans.). London: Routledge and Kegan Paul, Ltd., 1959.

—— (1924). *Judgment and Reasoning in the Child* (M. Warden, trans.). Totowa, N.J.: Littlefield, Adams and Co., 1972.

—— (1926). *The Child's Conception of the World* (J. and A. Tomlinson, trans.). Paterson, N. J.: Littlefield, Adams and Co., 1963.

—— (1932). *The Moral Judgment of the Child* (M. Gabain, trans.). New York: Free Press, 1965.

—— (1936a). *The Origins of Intelligence in Children* (M. Cook, trans.). New York: International Universities Press, 1974.

—— (1936b). *The Construction of Reality in the Child* (M. Cook, trans.). New York: Ballantine Books, 1954.

—— (1946). *Play, Dreams and Imitation in Childhood* (C. Gattegno and F. M. Hodgson, trans.). New York: W. W. Norton & Co., Inc., 1962.

—— (1947). *The Psychology of Intelligence* (M. Piercy and D. E. Berlyne, trans.). Totowa, N. J.: Littlefield, Adams and Co., 1973.

—— (1952). Autobiography. In E. Boring, H. S. Langfeld, H. Werner, and R. M. Yerkes, *A History of Psychology in Autobiography* (Vol. IV). Worcester, Mass.: Clark University Press.

—— (1964a). *Six Psychological Studies* (A. Tenzer and D. Elkind, trans.). New York: Vintage Books, 1968.

—— (1946b). Development and learning. In R. Ripple and V. Rockcastle (Eds.), *Piaget Rediscovered*. Ithaca, N.Y.: Cornell University Press, 1969.

—— (1969). *Science of Education and the Psychology of the Child* (D. Coltman, trans.). New York: Viking, 1970.

—— (1972). Intellectual evolution from adolescence to adulthood. *Human Development, 15,* 1-12.

——, and INHELDER, B. (1948). *The Child's Conception of Space* (F. J. Langdor and J. L. Lunzer, trans.). London: Routledge & Kegan Paul Ltd., 1956.

——, and INHELDER, B. (1966). *The Psychology of the Child* (H. Weaver, trans.). New York: Basic Books, 1969.

——, and SZEMINSKA, A. (1941). *The Child's Conception of Number* (C. Cattegno and F. M. Hodgson, trans.). New York: W. W. Norton & Co., Inc.

PODD, M. (1972). Ego identity status and morality. *Developmental Psychology, 6,* 497-507.

PREMACK, D. (1961). Predicting instrumental performance from the independent rate of the contingent response. *Journal of Experimental Psychology, 61,* 161-71.

REDL, F., and WINEMAN, D. (1951). *Children Who Hate.* New York: The Free Press of Glencoe.

REST, J. (1973). The hierarchical nature of moral judgment: The study of patterns of preference and comprehension of moral judgments made by others. *Journal of Personality, 41,* 86-109.

REYNOLDS, C. S. (1968). *A Primer of Operant Conditioning.* Glenview, Ill.: Scott, Foresman and Co.

RHEINGOLD, H. L., GEWIRTZ, J. L., and ROSS, H. W. (1959). Social conditioning of vocalizations in the infant. *Journal of Comparative and Physiological Psychology, 52,* 68-73.

RIESS, B. F. (1954). Effect of altered environment and of age on the mother-young relationships among animals. *Annals of the New York Academy of Science, 57,* 606-10.

RIMLAND, B. (1964). *Infantile Autism: The Syndrome and Its Implications for a Neural Theory of Behavior.* Englewood Cliffs, N.J.: Prentice-Hall.

ROSENTHAL, T. L. (1976). Modeling therapies. In M. Herson, R. M. Eisler, and P. M. Miller (Eds.), *Progress in Behavior Modification* (Vol. II). New York: Academic Press.

——, and ZIMMERMAN, B. J. (1972). Moderling by exemplification and instruction in training conservation. *Developmental Psychology, 6,* 392-401.

ROUSSEAU, J. J. (1750). Discourse on the Sciences and Arts. In R. D. Masters (Ed.), *The First and Second Discourses* (R. D. and J. R. Masters, trans.). New York: St. Martin's Press, 1964.

—— (1754). Discourse on the Origin and Foundations of Inequality. In R. D. Masters (Ed.), *The First and Second Discourses* (R. D. and J. R. Masters, trans.). New York: St. Martin's Press, 1964.

—— (1762a). *The Social Contract* (G. Hopkins, trans.). New York: Oxford University Press, 1962.

—— (1762b). *Emile, Or Education* (B. Foxley, trans.). London: J. M. Dent and Sons Ltd., 1948.

—— (1788). *The Confessions of Jean Jacques Rousseau.* New York: The Modern Library, 1945.

RUSHTON, J. P. (1975). Generosity in children: Immediate and long term effects of modeling, preaching, and moral judgment. *Journal of Personality and Social Psychology, 31,* 459-66.

RUSSELL, B. (1945). *A History of Western Philosophy.* New York: Simon & Schuster.

—— (1971). *Education and the Social Order.* London: George Allen and Unvin, Ltd.

SACHS, J. S. (1976). Development of speech. In E. C. Carterette, and M. P. Friedman (Eds.), *Handbook of Perception* (Vol. VII). New York: Academic Press.

SAHAKIAN, W. S., and SAHAKIAN, M. L. (1975). *John Locke.* Boston: Twayne Publishers.

de SANTIALLANA, G. (Ed.) (1956). *The Age of Adventure: The Renaissance Philosophers.* New York: Mentor.

SCHACHTEL, E. G. (1959). *Metamorphosis.* New York: Basic Books.

SCHNEIRLA, T. C. (1960). Instinctive behavior, maturation-experience, and development. In B. Kaplan and S. Wapner (Eds.), *Perspectives in Psychological Theory.* New York: International Universities Press.

SCHULTZ, D. P. (1975). *A History of Modern Psychology* (2nd ed.). New York: Academic Press.

—— (1976). *Theories of Personality.* Belmont, Calif.: Brooks/Cole.

SHEEHY, G. (1976). *Passages: Predictable Crises of Adult Life.* New York: Dutton.

SIGEL, I. E. (1968). Reflections. In I. E. Sigel, and F. H. Hooper (Eds.), *Logical Thinking in Children: Research Based on Piaget's Theory.* New York: Holt, Rinehard & Winston.

SINCLAIR, H. (1971). Sensorimotor action patterns as a condition for the acquisition of syntax. In R. Huxley and E. Ingram (Eds.), *Language Acquisition: Methods and Models.* New York: Academic Press.

SINGER, J. L. and SINGER, D. G. (1979). Come back, Mr. Rogers, come back. *Psychology Today, 12,* 56-60.

SKINNER, B. F. (1938). *The Behavior of Organisms.* Englewood Cliffs, N.J.: Prentice-Hall.

—— (1948). *Walden Two.* New York: Macmillan.

—— (1953). *Science and Human Behavior.* New York: Macmillan.

—— (1957). *Verbal Behavior.* Englewood Cliffs, N.J.: Prentice-Hall.

—— (1959). *Cumulative Record.* Englewood Cliffs, N.J.: Prentice-Hall.

—— (1967). Autobiography. In E. G. Boring, and G. Lindzey (Eds.), *A History of Psychology in Autobiography* (Vol. V). Englewood Cliffs, N.J.: Prentice-Hall.

—— (1968). *The Technology of Teaching.* Englewood Cliffs, N.J.: Prentice-Hall.

—— (1969). *Contingencies of Reinforcement.* Englewood Cliffs, N.J.: Prentice-Hall.

—— (1971). *Beyond Freedom and Dignity.* New York: Bantam.

—— (1974). *About Behaviorism.* New York: Knopf.

SLOBIN, D. I. (1971). *Psycholinguistics.* Glenview, Ill.: Scott, Foresman.

—— (1972). They learn the same way all around the world. *Psychology Today, 6,* 71-82.

—— (1973). Cognitive prerequisites for the development of grammar. In C. A. Ferguson and D. I. Slobin (Eds.), *Studies of Child Language Development.* New York: Holt, Rinehart & Winston.

SPOCK, B. (1945). *Baby and Child Care.* New York: Pocket Books, 1968.

SULLIVAN, H. S. (1953). *The Interpersonal Theory of Psychiatry.* New York: W. W. Norton & Co., Inc.

SULLIVAN, M. W. (1969). Programmed learning in reading. In A. D. Calvin (Ed.), *Programmed Instruction: Bold New Venture.* Bloomington, Ind.: Indiana University Press.

SZE, W. C. (Ed.) (1975). *Human Life Cycle.* New York: Jason Aronson.

TANNER, J. M. and INHELDER, B. (Eds.) (1971). *Discussions on Child Development.* The Proceedings of the Meetings of the World Health Organization Study Group on the Psychological Development of the Child, Geneva, 1953-56. New York: International Universities Press.

THOMPSON, C. (1950). Cultural pressures in the psychology of women. In P. Mullahy (Ed.), *A Study of Interpersonal Relations.* New York: Hermitage Press.

THORNDIKE, E. L. (1905). *The Elements of Psychology.* New York: Seiler.

TINBERGEN, N. (1951). *The Study of Instinct.* Oxford: Clarendon Press.

—— (1965). The shell menace. In T. E. McGill (Ed.), *Readings in Animal Behavior.* New York: Holt, Rinehart & Winston.

TOLMAN, E. C. (1948). Cognitive maps in rats and man. *Psychological Review, 55,* 189-208.

TREFFERT, D. A. (1970). Epidemiology of infantile autism. *Archives of General Psychiatry, 22,* 431-38.

TULKIN, S. R., and KONNER, M. J. (1973). Alternative conceptions of intellectual functioning. *Human Development, 16,* 33-52.

TURIEL, E. (1966). An experimental test of the sequentiality of developmental stages in the child's moral judgments. *Journal of Personality and Social Psychology, 3,* 611-18.

TURNER, E. W., and ROMMETVEIT, R. (1967). Experimental manipulation of the production of active and passive voice in children. *Language and Speech, 10,* 169-80.

UZGIRIS, I. C. (1964). Situational generality of conservation. *Child Development, 35,* 831-41.

VON FRANZ, M. L. (1964). The process of individuation. In C. G. Jung (Ed.), *Man and His Symbols.* New York: Dell Pub. Co., Inc.

WADDINGTON, C. H. (1962). *Biology for the Modern World.* New York: Barnes & Noble.

WATSON, J. B. (1913). Psychology as the behaviorist views it. *Psychological Review, 20,* 158-77.

—— (1924). *Behaviorism.* New York: W. W. Norton & Co., Inc., 1970.

—— (1928). *Psychological Care of Infant and Child.* New York: W. W. Norton & Co., Inc.

—— (1936). Autobiography. In C. Murchison (Ed.), *A History of Psychology in Autobiography* (Vol. III). Worcester, Mass.: Clark University Press.

WATSON, R. I. (1968). *The Great Psychologists from Aristotle to Freud* (2nd ed.). Philadelphia: Lippincott.

WERNER, H. (1948). *Comparative Psychology of Mental Development.* New York: Science Editions, 1965.

WHITE, G. M. (1972). Immediate and deferred effects of model observation and guided and unguided rehearsal on donating and stealing. *Journal of Personality and Social Psychology, 21* 139-48.

WHITE, R. W. (1960). Competence and the psychosexual stages of development. In M. Jones (Ed.), *Nebraska Symposium on Motivation.* Lincoln, Neb.: University of Nebraska Press.

——— (1963). Sense of interpersonal competence: Two case studies and some reflections on origins. In R. W. White (Ed.), *The Study of Lives.* New York: Atherton Press.

———, and WATT, N. F. (1973). *The Abnormal Personality* (4th ed.). New York: Ronald Press.

WHITE, S. (1965). Evidence for a hierarchical arrangement of learning processes. In L. P. Lipsitt, and C. C. Spiker (Eds.), *Advances in Child Development and Behavior* (Vol. II). New York: Academic Press.

WHITEHURST, G. J., and VASTA, R. (1975). Is language acquired through imitation? *Journal of Psycholinguistic Research, 4,* 37-59.

WHITING, J. W. M., and CHILD, I. L. (1953). *Child Training and Personality: A Cross-cultural Study.* New Haven, Conn.: Yale University Press.

WHITMONT, E. C. (1969). *The Symbolic Quest: Basic Concepts of Analytical Psychology.* New York: Putnam's.

———, and KAUFMANN, Y. (1973). Analytic psychotherapy. In R. Corsini (Ed.), *Current Psychotherapies.* Ithaca, Ill.: F. E. Peacock Publishers.

WILLIAMS, C. D. (1959). The elimination of tantrum behavior by extinction procedures. *Journal of Abnormal and Social Psychology, 5,* 269.

WILNER, W. (1975). Schachtel: A life. *William Alanson White Newsletter,* Winter, 3-4.

WILSON, C. (1972). *New Pathways in Psychology: Maslow and the Post-Freudian Revolution.* New York: Mentor.

WOLPE, J. (1969). *The Practice of Behavior Therapy.* New York: Pergamon Press.

ZIMMERMAN, B. J., and ROSENTHAL, T. L. (1974). Conserving and retaining equalities and inequalities through observation and correction. *Developmental Psychology, 10,* 260-68.

Name Index

Subject Index

293